Sustainable Civil Infrastructures

Editor-in-Chief

Hany Farouk Shehata, SSIGE, Soil-Interaction Group in Egypt SSIGE, Cairo, Egypt

Advisory Editors

Khalid M. ElZahaby, Housing and Building National Research Center, Giza, Egypt
Dar Hao Chen, Austin, TX, USA

Sustainable Infrastructure impacts our well-being and day-to-day lives. The infrastructures we are building today will shape our lives tomorrow. The complex and diverse nature of the impacts due to weather extremes on transportation and civil infrastructures can be seen in our roadways, bridges, and buildings. Extreme summer temperatures, droughts, flash floods, and rising numbers of freeze-thaw cycles pose challenges for civil infrastructure and can endanger public safety. We constantly hear how civil infrastructures need constant attention, preservation, and upgrading. Such improvements and developments would obviously benefit from our desired book series that provide sustainable engineering materials and designs. The economic impact is huge and much research has been conducted worldwide. The future holds many opportunities, not only for researchers in a given country, but also for the worldwide field engineers who apply and implement these technologies. We believe that no approach can succeed if it does not unite the efforts of various engineering disciplines from all over the world under one umbrella to offer a beacon of modern solutions to the global infrastructure. Experts from the various engineering disciplines around the globe will participate in this series, including: Geotechnical, Geological, Geoscience, Petroleum, Structural, Transportation, Bridge, Infrastructure, Energy, Architectural, Chemical and Materials, and other related Engineering disciplines.

More information about this series at http://www.springer.com/series/15140

Adam Bezvijen · Walter Wittke ·
Harry Poulos · Hany Shehata
Editors

Latest Advancements in Underground Structures and Geological Engineering

Proceedings of the 3rd GeoMEast
International Congress and Exhibition, Egypt
2019 on Sustainable Civil Infrastructures –
The Official International Congress
of the Soil-Structure Interaction Group
in Egypt (SSIGE)

Springer

Editors
Adam Bezvijen
Ghent University
Delft, The Netherlands

Harry Poulos
University of Sydney
Sydney, NSW, Australia

Walter Wittke
Global Expert of Rock Mechanics
and Tunnel
Munich, Germany

Hany Shehata
Soil-Structure Interaction
Group in Egypt (SSIGE)
Cairo, Egypt

ISSN 2366-3405 ISSN 2366-3413 (electronic)
Sustainable Civil Infrastructures
ISBN 978-3-030-34177-0 ISBN 978-3-030-34178-7 (eBook)
https://doi.org/10.1007/978-3-030-34178-7

This Springer imprint is published by the registered company Springer Nature Switzerland AG
The registered company address is: Gewerbestrasse 11, 6330 Cham, Switzerland

Contents

About the Editors

Adam Bezvijen is a professor in soil mechanics and geotechnics at Ghent University in Belgium and part-time senior specialist at Deltares, Delft, the Netherlands.

He worked full-time at Deltares before becoming a professor. At Deltares he was involved in research on revetments, dredging, tunnelling, geotextiles, and model testing and was the scientific coordinator of the geotechnical centrifuge of Deltares. As a professor in Ghent, he guides PhD and Post-doc research on tunnelling, piled embankments, geotextile reinforcement, polymer-treated bentonite, and backward erosion piping.

He is chair of the ISSMGE technical committee TC204 "Underground construction in Soft ground" and member of TC104 "Physical Model Testing".

Walter Wittke is a visionary and a devoted engineer. The Emeritus Professor and Founder of the engineering company WBI lives for his family and for his profession. Educated as a structural engineer, he dedicated his professional life to geotechnical and rock mechanical engineering. As one of the pioneers in rock mechanics, back in the 1970s, he started to develop models for the realistic description of the behaviour of jointed rock masses under consideration of discontinuity characteristics and implemented these in numerical codes. These models and codes form the basis for the realistic consideration of the interaction of ground and structure, which is of utmost importance in geotechnical works. Walter Wittke always put particular emphasis on the combination of research and practice. He successfully applied the technical models in numerous projects, refined, and expanded them. The result of decades of research and practice in rock mechanics is summarised in the book "Rock Mechanics Based on an Anisotropic Jointed Rock Model (AJRM)", which was published in 2004. After long years of serving the geotechnical community and the university, Walter Wittke is now focusing on his work as an engineer in his company WBI, together with his children and his employees, and thus continues to indulge his passion for geotechnical engineering.

Professor Dr. Harry Poulos
AM FAA FTSE Hon FIEAust Dist M ASCE

Professor Harry Poulos' pioneering work in pile foundation analysis and design has enabled the world's geotechnical specialists to have a greater understanding of the way structures interact with the ground. His research has enabled a more reliable approach to be adopted for pile design, replacing procedures which previously relied purely on experience and empiricism.

Professor Poulos has applied his research to a wide range of major projects, both in Australia and overseas, including buildings, bridges, tunnels, freeways, mines, airports offshore structures (e.g. oil rigs) and earthquake-related problems. Professor Poulos' work includes the Emirates Twin Towers in Dubai, where his analysis and design of the piled raft foundations provided significant cost benefits for the Twin Towers exceeding 300 metres in height, the Burj Khalifa, now the world's tallest building, where he was the geotechnical peer reviewer, the Docklands project in Melbourne involving design of remedial pile foundations for one of the high rise residential developments, and the construction of a 700 km long motorway in Greece using his expertise in slope stabilisation and earthquake engineering.

While retaining his professorial position at the University of Sydney, Professor Poulos joined the Coffey Group in 1989, as the Director of Advanced Technology, and became Chairman of Coffey International Pty Ltd in 1991, a position that he held for two years. In the period 1998 to 2002, he served as Director of Technical Innovation and General Manager, Technical Development.

Professor Poulos has long been a contributor to the activities of the international geotechnical community. He was also a long-term member of the National Committee of the Australian Geomechanics Society (AGS) (1980 to 1995) and its Chairman from 1982 to 1984. He was Committee Member of the AGS Sydney Group, 1971–1976, 1979–2002, Vice-chairman 1974 and Chairman 1980–1981. He was the Australasian Vice-President of the International Society for Soil Mechanics and Foundation Engineering in the period 1989–1994, an appointed Board Member of the Society from 2001 to 2005, and is currently the Chair of the Membership, Practitioner and Academic Committee of the Society.

He was recognised by his peers for his contributions to Australian Geomechanics by the Sydney Chapter via the institution of the annual Poulos Lecture in 2002.

Professor Poulos is a recipient of many prizes and awards, including Australia's Centenary Medal (2003) for his services to Australian society and science in the field of geotechnical engineering. His overall contribution to the engineering profession has been recognised formally by the award of Member of the Order of Australia (1993), his election as Fellow of the Australian Academy of Science (1988), his Fellowship of the Australian Academy of Technological Sciences and Engineering (1996), his Honorary Fellowship of the Institution of Engineers Australia (1999), the award of the Warren Prize (1972) and Warren Medal (1985) of the Institution of Engineers, Australia, his selection as the 2003 Australian Civil Engineer of the year, and his selection in 2004 as the inaugural Geotechnical Practitioner of the year.

Professor Poulos gave the prestigious Rankine Lecture of the Institution of Civil Engineers (UK) in 1989 and was invited by the American Society of Civil Engineers (ASCE) to deliver the annual Terzaghi lecture in 2004. He also received from ASCE the 1972 Croes Medal, the 1995 State-of-the-Art Award, and the 2007 Middlebrooks Award. In 2010, he was elected as a Distinguished Member of ASCE, the first Australian Civil Engineer to be so recognised.

Hany Shehata is the founder and CEO of the Soil-Structure Interaction Group in Egypt "SSIGE". He is a partner and Vice-President of EHE-Consulting Group in the Middle East and managing editor of the "Innovative Infrastructure Solutions" journal, published by Springer. He worked in the field of civil engineering early, while studying, with Bechtel Egypt Contracting & PM Company, LLC. His professional experience includes working in culverts, small tunnels, pipe installation, earth reinforcement, soil stabilization, and small bridges. He also has been involved in teaching, research, and consulting. His areas of specialisation include static and dynamic soil-structure interactions involving buildings, roads, water structures, retaining walls, earth reinforcement, and bridges, as well as, different disciplines of project management and contract administration. He is the author of an Arabic practical book titled "Practical Solutions for Different Geotechnical Works: The Practical Engineers' Guidelines". He is currently working on a new book titled "Soil-Foundation-Superstructure Interaction: Structural Integration". He is the contributor of more than 50 publications in national and international conferences and journals. He served as a co-chair of the GeoChina 2016 International Conference in Shandong, China. He serves also as a co-chair and secretary general of the GeoMEast 2017 International Conference in Sharm El-Sheikh, Egypt. He received the Outstanding Reviewer of the ASCE for 2016 as selected by the Editorial Board of International Journal of Geomechanics.

Defining the Contours of Combined Stressed, Anomaly Elastic and Anomaly Densed Hierarchical Inclusions Located into a Block Layered Medium by Wave's Data of Active Acoustic and Electromagnetic Monitoring

Olga Hachay[1]([✉]), Andrey Khachay[2], and Oleg Khachay[2]

[1] Institute of Geophysics Ural Branch of Russian Academy of Sciences,
Yekaterinburg, Russia
olgakhachay@yandex.ru
[2] Ural Federal University, Yekaterinburg, Russia

Abstract. A new approach to the interpretation of wave fields has been developed to determine the contours or surfaces of composite local hierarchical objects. An iterative process has been developed to solve a theoretical inverse problem for the case of determining the configurations of 2D hierarchical inclusions of the l-th, m-th, and s-th ranks located one above the other in different layers of the N-layer medium and various physical and mechanical properties for active acoustic monitoring with sources of longitudinal and transversal waves. When interpreting the results of monitoring, it is necessary to use data from such observation systems that can be configured to study the hierarchical structure of the environment. Such systems include acoustic (in the dynamic version) and electromagnetic monitoring systems. The hierarchical structure of the geological environment is clearly visible when analyzing rock samples taken from ore mines. On the other hand, the more complex the environment, the each wave field introduces its information about its internal structure, therefore, the interpretation of the seismic and electromagnetic fields must be conducted separately, without mixing these databases. This result is contained in the explicit form of the equations of the theoretical inverse problem for a 2D electromagnetic field (E and H polarization), as well as for the propagation of a linearly polarized elastic wave when excited by an N-layer conducting or elastic medium with a hierarchical conducting or elastic inclusion located in the v-th layer. In the present work, the inverse problem for a complicated hierarchical model of inclusions is considered. It can be used when conducting monitoring seismic and acoustic borehole studies to monitor the fluid return of oil fields, to analyze the dynamic state of a mountain range of deep-seated deposits which are under various mechanical effects.

Keywords: Combined hierarchical environment · Acoustic and electromagnetic field · Iterative algorithm · Equation of the theoretical inverse problem

A. Bezvijen et al. (Eds.): GeoMEast 2019, SUCI, pp. 1–11, 2020.
https://doi.org/10.1007/978-3-030-34178-7_1

An important role for understanding the formation and development of a hierarchy of structural levels of deformation in solids is played by theoretical and experimental results obtained on samples (Panin et al. 1985). With their help, an approach has been developed that uses ideas about dissipative structures in noneqiulibrium systems (Nikolis and Prigozin 1979), for which self-organization processes take place at each of the hierarchical levels. As shown in Nikolis (1989), self-organization occurs when there is a hierarchical structure. This approach can be applied to the study of such natural and man-made systems, such as rock massifs, which are in the process of mining. The model of an open dynamic system (Nikolis and Prigozin 1990) is applicable for their description. Analysis of the manifestations of self-organization processes can give an idea of the stability of the system and contribute to the development of criteria for the stability of the state of the array as a whole regarding the dynamic phenomena of a given energy class. This satisfies the statement expressed in (Goldin 2002), which consists in the hypothesis about the divisibility of medium scales. While the destruction of smaller scales fits into the concept of a nonstationary random process, for which the prediction of individual events is not possible.

In Hachay (2004), Hachay et al. (2003), using 3D electromagnetic induction space-time monitoring (Hachay 2007; Hachay et al. 2001), it was possible to show that the model of a hierarchical discrete environment is applicable to describe the structure of an array of rocks of different material composition. Within the framework of a specific modification of the method, it was possible to trace two hierarchical levels. The zones of disintegration (Shemjakin et al. 1986; Shemjakin et al. 1992) in the near-working space are located asymmetrically in the soil and roof, which may be evidence of a non-equilibrium state of the system. These zones are located discretely, i.e. there are intervals of their total absence in the near-working space. The maximum changes in the massif, which is under the anthropogenic influence, manifest themselves in the change over time of the morphology of the spatial position of these zones (Fig. 1 (a–d)).

At present, theoretical results on modeling the electromagnetic and seismic field in a layered medium with inclusions of a hierarchical structure are in demand. Simulation algorithms are constructed in the electromagnetic case for the 3D heterogeneity, in the seismic case for the 2D heterogeneity (Hachay and Khachay 2016a; Hachay and Khachay 2017; Hachay et al 2018a, b; Hachay et al. 2015; Hachay et al. 2016; Hachay et al. 2017a). It is shown, that with the increase of hierarchy degree of the environment, the degree of spatial distribution nonlinearity of the seismic and electromagnetic fields components increases also. That corresponds to the detailed monitoring experiments conducted in the hazardous mines of the Tashtagol mine and SUBR. The theory developed demonstrated how complex the process of integrating methods using an electromagnetic and seismic field to study the response of a medium with a hierarchical structure. This problem is inextricably linked with the formulation and solution of the inverse problem for the propagation of electromagnetic and seismic fields in such complex environments. In Hachay and Khachay (2015), Hachay and Khachay (2016b), Hachay et al. (2017b), the problem of constructing an algorithm for solving an inverse problem using the equation of a theoretical inverse problem for the 2D Helmholtz equation was considered. Explicit equations of the theoretical inverse problem are written for cases of electromagnetic field scattering (E and H polarization) and scattering of a linearly polarized elastic wave in a layered conducting and elastic medium

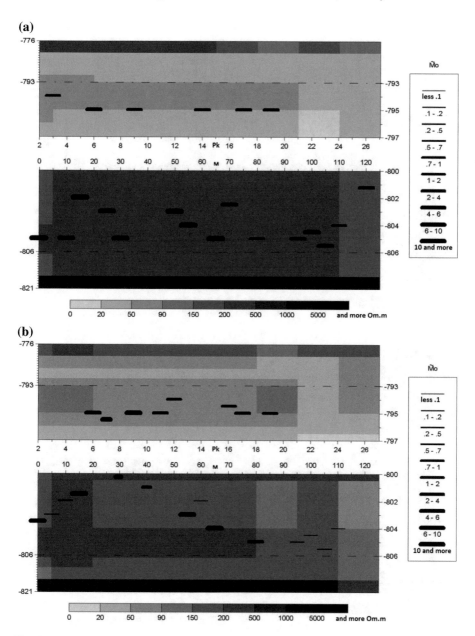

Fig. 1. The manifestation of the process of self-organization in the morphology of the zones of disintegration identified according to the data of electromagnetic induction monitoring (a) Geoelectrical section according to ort 19, horizon −350, frequency 20 kHz, observations 2002; (b) Geoelectrical section of the ort 19, horizon −350, frequency 20 kHz, observations 2003; (c) Geoelectrical section ort 8, horizon −210, frequency 10 kHz, observations 2002; (d) Geoelectrical section in line 8, horizon −210, frequency 10 kHz, observations 2003.

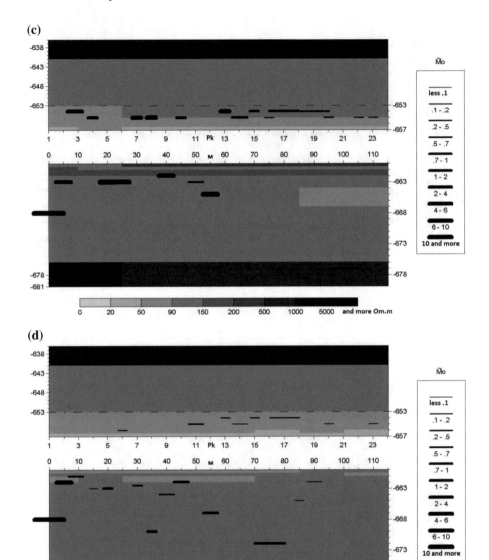

Fig. 1. (*continued*)

with a hierarchical conducting or elastic inclusion, which are the basis for determining the contours of misaligned inclusions of the l-th rank of a hierarchical structure. Obviously, when solving the inverse problem, it is necessary to use observation systems set up to study the hierarchical structure of the environment as the initial monitoring data. In Hachay et al. (2018a), a modeling algorithm was constructed for the acoustic monitoring data of a hierarchical two-phase geological environment with different physical and mechanical properties. In this paper, we construct an algorithm for reconstructing the contours of hierarchical composite structures associated with disintegration zones according to active acoustic monitoring using a source of longitudinal waves.

1 Algorithm for Solution of the Inverse Problem of 2-D Sound Diffraction in N-Layered Medium with Composite Hierarchical Inclusions

The inverse problem of the diffraction of a linearly polarized elastic shear wave by a two-dimensional elastic heterogeneity of a hierarchical type located in a layer v of an N-layered medium was solved in Hachay et al. (2017b). Here we consider this problem for a source of a longitudinal wave in the framework of a complicated model: the anomalously stressed hierarchical heterogeneity of the l-th rank will be located in the layer $v - 1$, the maximum value of the l-th rank is L, the initial value of the l-th rank is ll = 1; the anomalously elastic hierarchical heterogeneity of the m-th rank is located in the layer v, the maximum value of the m-th rank is M, the initial value of the m-th rank is mm = 1, and the anomalously densed hierarchical heterogeneity of the s-th rank is located in the layer $v + 1$, the maximum value of the s-th rank is equal to S, the initial value of the s-th rank is ss = 1. We consider the algorithm for recovering 2D surfaces of hierarchical heterogeneities in the case when L < M < S. Let us write the equation of the theoretical inverse problem (Hachay et al. 2017b) for the scalar Helmholtz equation, to which our problem for the layer $v - 1$ reduces:

$$2\pi U^+ (M_0) = \int_{\partial D} ((U_{v-1}^+(M) + U_{v-1}^1(M))(\frac{\partial G^a(M, M_0)}{\partial n} - (b_{v-1}/b_i)\frac{\partial G(M, M_0)}{\partial n})$$

$$- b_{v-1}(\frac{\partial U_{v-1}^+}{\partial n} + \frac{\partial U_{v-1}^1}{\partial n})((1/b_a)G^a(M, M_0) - (1/b_i)G(M, M_0)))dl; \tag{1}$$

$$\text{By this } b_{v-1} = \xi_{v-1}; b_i = \xi_i; b_{a(v-1)l} = \xi_{a(v-1)l}; l = ll \tag{2}$$

$\xi_{v-1}, \xi_i, \xi_{a(v-1)l}, \rho_{v-1}, \rho_i, \rho_{a(v-1)l}$- the values of the elastic parameter Lamé λ and the density in the $(v - 1)$ -th layer, in the layer where point M_0 is located inside the heterogeneity in the layer, with:

$$\xi_{a(v-1)l} = \lambda_{(v-1)al}; \xi_{v-1} = \lambda_{v-1}; \xi_i = \lambda_i; \rho_{v-1} = \rho_{a(v-1)l};$$

$$U^+ = \varphi^+; U_{v-1}^+ = \varphi_{(v-1)}^+; U_{v-1}^1 = \varphi_{(v-1)}^1 \tag{3}$$

Where $\vec{u} = grad\varphi$, if l = 1, then $\varphi_{(v-1)}^1$ is equal to the potential by absence of heterogeneities in the layered medium; if l > 1, then $\varphi_{(v-1)}^1 = \varphi_{(v-1)}$ (see the value calculated by formula (21).

$$G(M, M_0) = G_{SP}(M, M_0); G^a(M, M) = G^a_{SP}(M, M_0); \partial D, dL,$$

$$k^2_{1a(v-1)l} = \omega^2 \frac{\rho_{a(v-1)l}}{\xi_{a(v-1)l}}; \ k^2_{1(v-1)} = \omega^2 \frac{\rho_{(v-1)}}{\xi_{(v-1)}}; k^2_{1i} = \omega^2 \frac{\rho_i}{\xi_i}; \tag{4}$$

The algorithm for calculating the Green function is written in Khachay (2016a, b). Thus, the equation of the theoretical inverse problem is written in the form:

$$2\pi\varphi^+_{(v-1)l}(M_0) = \int_{\partial Dl} ((\varphi^+_{(v-1)l}(M) + \varphi^1_{(v-1)}(M))(\frac{\partial G^{al}_{SP}(M, M_0)}{\partial n} - (\xi_{(v-1)}/\xi_i)\frac{\partial G_{SP}(M, M_0)}{\partial n})$$

$$- \xi_{(v-1)}(\frac{\partial \varphi^+_{(v-1)l}}{\partial n} + \frac{\partial \varphi^1_{(v-1)}}{\partial n})((1/\xi_{a(v-1)l})G^{al}_{SP}(M, M_0) - (1/\xi_i)G_{SP}(M, M_0)))dL;$$

$$\tag{5}$$

Solving Eq. (5) with respect to the function describing the contour, we calculate the functions: $\varphi_{(v-1)}$; φ_v; $\varphi^+_{(v-1)l}$; $\varphi^1_{(v-1)}$ by the algorithm for solving the direct problem (Hachay et al. 2017a) inside and outside the heterogeneity placed in a layered medium.

$$\frac{(k^2_{1(v-1)l} - k^2_{1(v-1)})}{2\pi} \iint_{S1cl} \varphi_l(M)G_{SP,(v-1)}(M, M^0)d\tau_M + \varphi^1_{l-1}(M^0) = \varphi_l(M^0), M^0 \in S_{1Cl}$$

$$\frac{\rho_{(v-1)l}(k^2_{1(v-1)l} - k^2_{1(v-1)})}{\rho(M^0)2\pi} \iint_{S1cl} \varphi_l(M)G_{SP,(v-1)}(M, M^0)d\tau_M + \varphi^1_{l-1}(M^0) = \varphi_l(M^0), M^0 \notin S_{1Cl}, M^0 \in \Pi_{v-1}$$

$$\tag{6}$$

$$\frac{\rho_{(v)l}(k^2_{1(v)l} - k^2_{1(v)})}{\rho(M^0)2\pi} \iint_{S1cl} \varphi_l(M)G_{SP,(v)}(M, M^0)d\tau_M + \varphi^1_{l-1}(M^0) = \varphi_l(M^0), M^0 \notin S_{1Cl} M^0 \in \Pi_v$$

$$\tag{7}$$

Π_v- region of the layer v, Π_{v-1}- region of the layer v − 1.

Let us write the equation of the theoretical inverse problem (Hachay et al. 2017b) for the scalar Helmholtz equation, to which our problem for the layer v reduces, m = mm:

$$2\pi U^+(M_0) = \int_{\partial D} ((U^+_v(M) + U^1_v(M))(\frac{\partial G^a(M, M_0)}{\partial n} - (b_v/b_i)\frac{\partial G(M, M_0)}{\partial n})$$

$$- b_v(\frac{\partial U^+_v}{\partial n} + \frac{\partial U^1_v}{\partial n})((1/b_a)G^a(M, M_0) - (1/b_i)G(M, M_0)))dL; \tag{8}$$

By that $b_v = \xi_v; b_i = \xi_i; b_{a(v)m} = \xi_{a(v)m}$; $\tag{9}$

$\xi_v, \xi_i, \xi_{a(v)m}, \rho_v, \rho_i, \rho_{a(v)m}$- values of the elastic parameter λ Lame and density in the (v) -th layer, in the layer where the point M_0 is located and inside the heterogeneity in the layer v, by that:

$$\xi_{a(v)m} = \lambda_{(v)am}; \xi_v = \lambda_v; \xi_i = \lambda_i; \xi_i = \lambda_i; \rho_v \neq \rho_{a(v)m}$$
$$U^+ = \varphi^+; U_v^+ = \varphi_{(v)}^+; U_v^1 = \varphi_{(v)}^1 \tag{10}$$

calculated by the formula (7);
where $\vec{u} = grad\varphi$, φ is the potential of the acoustic field excited by a longitudinal wave for the selected model.

$$G(M, M_0) = G_{SP}(M, M_0); G^a(M, M) = G_{SP}^a(M, M_0); \partial D, dL,$$
$$k_{1a(v)m}^2 = \omega^2 \frac{\rho_{a(v)m}}{\xi_{a(v)m}}; \quad k_{1(v)}^2 = \omega^2 \frac{\rho_{(v)}}{\xi_{(v)}}; k_{1i}^2 = \omega^2 \frac{\rho_i}{\xi_i}; \tag{11}$$

Thus, the equation of the theoretical inverse problem is written in the form:

$$2\pi\varphi_{(v)m}^+(M_0) = \int\limits_{\partial Dm} ((\varphi_{(v)m}^+(M) + \varphi_{(v)}^1(M))(\frac{\partial G_{SP}^{am}(M, M_0)}{\partial n} - (\xi_{(v)}/\xi_i)\frac{\partial G_{SP}(M, M_0)}{\partial n})$$
$$- \xi_{(v)}(\frac{\partial \varphi_{(v)m}^+}{\partial n} + \frac{\partial \varphi_{(v)}^1}{\partial n})((1/\xi_{a(v)m})G_{SP}^{am}(M, M_0) - (1/\xi_i)G_{SP}(M, M_0)))dL; \tag{12}$$

Solving the Eq. (12) with respect to the function $r_m(\varphi)$ describing the contour ∂D_m, we calculate the functions: $\varphi_{(v)}$; $\varphi_{(v)m}^+$; $\varphi_{(v)}^1$ by the algorithm for solving the direct problem (Hachay et al. 2018b) inside and outside the heterogeneity located in the layered medium.

$$\frac{(k_{1vm}^2 - k_{1v}^2)}{2\pi} \iint\limits_{S_{2Cm}} \varphi_m(M)G_{SP,v}(M, M^0)d\tau_M + \frac{\rho_{va}}{\rho_{vm}}\varphi_{m-1}^1(M^0) - \frac{(\rho_{va} - \rho_{vm})}{\rho_{vm}2\pi} \int\limits_{C_{2m}} G_{SP,v}\frac{\partial \varphi_m}{\partial n}dc$$
$$= \phi_m(M^0), M^0 \in S_{2Cm};$$
$$\frac{\rho_{vm}(k_{1vm}^2 - k_{1v}^2)}{\rho(M^0)2\pi} \iint\limits_{S_{2Cm}} \varphi_m(M)G_{SP,v}(M, M^0)d\tau_M + \varphi_{m-1}^1(M^0) - \frac{(\rho_{va} - \rho_{vm})}{\rho(M^0)2\pi} \int\limits_{C_{2m}} G_{SP,v}\frac{\partial \varphi_m}{\partial n}dc$$
$$= \varphi_m(M^0), M^0 \notin S_{2Cm} \tag{13}$$

$$\frac{\rho_{(v+1)m}(k_{1(v+1)m}^2 - k_{1(v+1)}^2)}{\rho(M^0)2\pi} \iint\limits_{S_{2Cm}} \varphi_m(M)G_{SP,v+1}(M, M^0)d\tau_M + \varphi_{m-1}^1(M^0)$$
$$- \frac{(\rho_{(v+1)a} - \rho_{(v+1)m})}{\rho(M^0)2\pi} \int\limits_{C_{2m}} G_{SP,v+1}\frac{\partial \varphi_m}{\partial n}dc = \varphi_m(M^0), M^0 \notin S_{2Cm}, \in \Pi_{v+1}; \tag{14}$$

We write the equation of the theoretical inverse problem (Hachay et al. 2017b) for the scalar Helmholtz equation, to which our problem reduces, for the layer $v + 1$, $s = ss$:

$$2\pi U^+(M_0) = \int\limits_{\partial D} ((U^+_{v+1}(M) + U^1_{v+1}(M))(\frac{\partial G^a(M,M_0)}{\partial n} - (b_{v+1}/b_i)\frac{\partial G(M,M_0)}{\partial n})$$

$$- b_{v+1}(\frac{\partial U^+_{v+1}}{\partial n} + \frac{\partial U^1_{v+1}}{\partial n})((1/b_a)G^a(M,M_0) - (1/b_i)G(M,M_0)))dL;$$

$$(15)$$

By that:

$$b_{v+1} = \xi_{v+1}; \; b_i = \xi_i; \; b_{a(v+1)s} = \xi_{a(v+1)s}; \qquad (16)$$

$\xi_v, \xi_i, \xi_{a(v)s}, \rho_v, \rho_i, \rho_{a(v)s}$- values of the elastic parameter Lame λ and density in the $(v + 1)$ -th layer, in the layer where point M_0 is located and inside the heterogeneity in the layer $(v + 1)$:

$$\xi_{a(v+1)s} = \lambda_{(v+1)as} = \lambda_i; \xi_{v+1} = \lambda_{v+1}; \xi_i = \lambda_i; \rho_i \neq \rho_{a(v+1)s}$$

$$U^+ = \varphi^+; \; U^+_{v+1} = \varphi^+_{(v+1)}; \; U^1_{v+1} = \varphi^1_{(v+1)} = \varphi_{(v+1)m} \qquad (17)$$

$$G(M,M_0) = G_{SP}(M,M_0); G^a(M,M) = G^a_{SP}(M,M_0); \partial D, dL,$$

$$k^2_{1a(v+1)s} = \omega^2\frac{\rho_{a(v+1)s}}{\xi_{a(v+1)s}}; \; k^2_{1(v+1)} = \omega^2\frac{\rho_{(v+1)}}{\xi_{(v+1)}}; k^2_{1i} = \omega^2\frac{\rho_i}{\xi_i}; \qquad (18)$$

Thus, the equation of the theoretical inverse problem is written in the form:

$$2\pi\varphi'^+_{(v+1)s}(M^0) = \int\limits_{\partial Ds} ((\varphi^+_{(v+1)s}(M) + \varphi^1_{(v+1)}(M))(\frac{\partial G^{as}_{SP}(M,M^0)}{\partial n} - (\xi_{(v+1)}/\xi_i)\frac{\partial G_{SP}(M,M^0)}{\partial n})$$

$$- \xi_{(v+1)}(\frac{\partial\varphi^+_{(v+1)s}}{\partial n} + \frac{\partial\phi^1_{(v+1)}}{\partial n})((1/\xi_{a(v+1)s})G^{as}_{SP}(M,M^0) - (1/\xi_i)G_{SP}(M,M^0)))dL;$$

$$(19)$$

Solving the Eq. (19) with respect to the function $r_s(\varphi)$ describing the contour ∂D_s, we calculate the functions: $\varphi_{(v+1)}; \varphi^+_{(v+1)s}; \varphi^1_{(v+1)}$ by the algorithm for solving the direct problem (Hachay et al. 2017a) inside and outside the heterogeneity placed in a layered medium.

$$\frac{(k^2_{1(v+1)is} - k^2_{1(v+1)})}{2\pi} \iint\limits_{S_{3Cs}} \varphi_s(M)G_{SP,(v+1)}(M,M^0)d\tau_M + \frac{\rho_{(v+1)a}}{\rho_{(v+1)is}}\varphi^1_{s-1}(M^0)$$

$$- \frac{(\rho_{(v+1)a} - \rho_{(v+1)is})}{\rho(M^0)2\pi}\int\limits_{C_{3s}} G_{SP,(v+1)}\frac{\partial\varphi_s}{\partial n}dc = \varphi_s(M^0), M^0 \in S_{3Cs}; \qquad (20)$$

If the rank $l \leq L$, $\varphi_{(v-1)}(M_0)$ is calculated by the formula (21):

$$
\frac{\rho_{(v-1)is}(k_{1(v-1)is}^2 - k_{1(v-1)}^2)}{\rho(M^0)2\pi} \iint\limits_{S_{3Cs}} \varphi_s(M) G_{SP,(v-1)}(M, M^0) d\tau_M + \varphi_{s-1}^1(M^0)
$$
$$
- \frac{(\rho_{(v-1)a} - \rho_{(v-1)is})}{\rho(M^0)2\pi} \int\limits_{C_m} G_{SP,(v-1)} \frac{\partial \varphi_s}{\partial n} dc = \varphi_{s(v-1)}(M^0), M^0 \notin S_{3Cs} \in \Pi_{v-1};
$$

(21)

ll = ll + 1 and we go to (1). If the rank is $l > L$, and $m \leq M$, then $\varphi_{(v)}(M_0)$ is calculated by the formula (22):

$$
\frac{\rho_{(v)is}(k_{1(v)is}^2 - k_{1(v)}^2)}{\rho(M^0)2\pi} \iint\limits_{S_{3Cs}} \varphi_s(M) G_{SP,(v)}(M, M^0) d\tau_M + \varphi_{s-1}^1(M^0)
$$
$$
- \frac{(\rho_{(v)a} - \rho_{(v)is})}{\rho(M^0)2\pi} \int\limits_{C_m} G_{SP,(v)} \frac{\partial \varphi_s}{\partial n} dc = \varphi_{s(v)}(M^0), M^0 \notin S_{3Cs} \in \Pi_v;
$$

(22)

mm = mm + 1 and we go to (8). If the rank is m > M, then $\varphi_{(v+1)}(M_0)$ is calculated by the formula (23):

$$
\frac{\rho_{(v+1)is}(k_{1(v+1)is}^2 - k_{1(v+1)}^2)}{\rho(M^0)2\pi} \iint\limits_{S_{3Cs}} \varphi_s(M) G_{SP,(v+1)}(M, M^0) d\tau_M + \varphi_{s-1}^1(M^0)
$$
$$
- \frac{(\rho_{(v+1)a} - \rho_{(v+1)is})}{\rho(M^0)2\pi} \int\limits_{C_m} G_{SP,(v+1)} \frac{\partial \varphi_s}{\partial n} dc = \varphi_{s(v+1)}(M^0), M^0 \notin S_{3Cs} \in \Pi_{v+1};
$$

(23)

ss = ss + 1 and we go to (15). If the rank is s > S, then $\phi_{(v+1)}(M_0)$ is calculated by the formula (24) in all layers: $v = 1, \ldots N$.

$$
\frac{\rho_{(v)is}(k_{1(v)is}^2 - k_{1(v)}^2)}{\rho(M^0)2\pi} \iint\limits_{S_{3Cs}} \varphi_s(M) G_{SP,(v)}(M, M^0) d\tau_M + \varphi_{s-1}^1(M^0)
$$
$$
- \frac{(\rho_{(v)a} - \rho_{(v)is})}{\rho(M^0)2\pi} \int\limits_{C_m} G_{SP,(v)} \frac{\partial \varphi_s}{\partial n} dc = \varphi_{s(v)}(M^0), M^0 \notin S_{3Cs} \in \Pi_v;
$$

(24)

The result is the defining of all the nested surfaces of hierarchical heterogeneities according to the ranks of their hierarchy and given physical and mechanical properties.

2 Conclusion

When solving the inverse problem, it is necessary to use observation systems set up to study the hierarchical structure of the environment as basic monitoring data. On the other hand, the more complex the environment, the each wave field introduces its information about its internal structure, therefore, the interpretation of the seismic and electromagnetic fields must be conducted separately, without mixing these databases. From the constructed theory it follows that with an increase in the degree of hierarchy of the environment, the degree of spatial nonlinearity of the distribution of the components of the seismic and electromagnetic fields increases, which indicates the impossibility of using the methods of linearizing the task when creating interpretation methods.

References

Goldin, S.V.: Lithosphere destruction and physical mesomechanics. Phys. Mesomech. **5**(5), 5–22 (2002)

Nikolis, G., Prigozin, I.: Self-Organization in Non-Equilibrium Systems. Mir, Moscow (1979)

Nikolis, G.: Dynamics of Hierarchical Systems. Mir, Moscow (1989)

Nikolis, G., Prigozin, I.: Knowledge of the Complex. Mir. Moscow (1990)

Panin, V.E., et al.: Structural levels of solids deformation. Nauka SB AN USSR, Novosibirsk (1985)

Hachay, O.A., et al.: Three-dimensional electromagnetic monitoring of the state of the rock massif. Earth Phys. **2**, 85–92 (2001)

Hachay, O.A., Novgorodova, E.N., Khachay, O.Yu.: A new technique for the detection of zones of disintegration in the near-working space of rock massifs. Min. Inf. Anal. Bull. **11**, 26–29 (2003)

Hachay, O.A.: To the question of studying the structure and state of a geological heterogeneous unsteady medium in the framework of a discrete hierarchical model. Russ. Geophys. J. **33–34**, 32–37 (2004)

Hachay, O.A.: Investigation of the development of instability in the rock mass using the method of active electromagnetic monitoring. Earth Phys. **4**, 65–70 (2007)

Hachay, O.A., Khachay, A.Yu.: Determination of the surface of fluid-saturated porous inclusion in a hierarchical layered-block medium according to electromagnetic monitoring data. Min. Inf. Anal. Bull. **4**, 150–154 (2015)

Hachay, O.A., Khachay, O.Yu., Khachay, A.Yu.: New methods of geoinformatic for monitoring wave fields in hierarchical environments. Geoinformatics **3**, 45–51 (2015)

Hachay, O.A., Khachay, A.Yu.: Simulation of seismic field propagation in a layered-block elastic medium with hierarchical plastic inclusions. Min. Inf. Anal. Bull. **12**, 318–326 (2016a)

Hachay, O.A., Khachay, A.Yu.: Determination of the surface of abnormally intense inclusion in a hierarchical layered-block medium according to acoustic monitoring data. Min. Inf. Anal. Bull. **4**, 354–356 (2016b)

Hachay, O.A., Khachay, O.Yu., Khachay, A.Yu.: New methods of geoinformatic for the integration of seismic and gravitational fields in hierarchical environments. Geoinformatics **3**, 25–29 (2016)

Hachay, O., Khachay, A.: Acoustic wave monitoring of fluid dynamic in the rock massif with anomaly of density, stressed and plastic hierarchic inclusions. Comput. Exp. Stud. Acoust. Waves. **4**, 63–80 (2017). http://dx.doi/org/10/5772/intechopen70590. InTech open. Chapter

Hachay, O.A., Khachay, O.Yu., Khachay, A.Yu.: Integration of acoustic, gravitational, and geomechanical fields in hierarchical environments. Min. Inf. Anal. Bull. **4**, 328–336 (2017a)

Hachay, O.A., Khachay, O.Yu., Khachay, A.Yu.: On the question of the inverse problem of active electromagnetic and acoustic monitoring of a hierarchical geological environment. Geophys. Res. **18**(4), 71–84 (2017b). https://doi.org/10.21455/gr2017/4-6

Hachay, O.A., Khachay, A.Yu., Khachay, O.Yu.: To the question of modeling acoustic monitoring of a hierarchical two-phase geological environment with different physical and mechanical properties. Monit. Sci. Technol. (1), 45–49 (2018a)

Hachay, O.A., Khachay, A.Yu., Khachay, O.Yu.: Modeling algorithm of acoustic waves penetrating through a medium with composite hierarchical inclusions. In: AIP Conference Proceedings, vol. 2053, p. 030023 (2018b). https://doi.org/10.1063/1.5084384

Khachay, A.Y.: Algorithm for solving the direct dynamic seismic problem when excited by a horizontal point force located in an arbitrary layer of an N-layer elastic isotropic medium. Inform. Math. Model. (2006a). USSU, Ekaterinburg

Khachay, A.Y.: Algorithm for solving the direct dynamic seismic problem when excited by a vertical point force located in an arbitrary layer of an N-layer elastic isotropic medium. Inform. Math. Model. (2006b). USSU, Ekaterinburg

Shemjakin, E.I., Fisenko, G.L., Kurlenja, M.V., Oparin, V.N., et al.: The effect of zonal disintegration of rocks around underground holes. DAN USSR **289**(5), 1088–1094 (1986)

Shemjakin, E.I., Kurlenja, M.V., Oparin, V.N., et al.: Discovery number 400. The phenomenon of rocks zonal disintegration around underground holes. Bull. Discov., 1 (1992)

A Closer Look: Petrographic Analysis of Extremely Weak Sandstone/Cemented Sand of the Ghayathi Formation, Dubai, UAE

Luke Bernhard Brouwers[✉]

Fugro Middle East, Al Quoz Industrial Area, 2863 Dubai, UAE
l.brouwers@fugro.com

Abstract. An imminent challenge currently being faced in Dubai is the occurrence of soft rock underlying the quaternary aeolian sand deposits. One such offender encountered, is the thinly laminated calcareous sandstone of the Ghayathi Formation, which may occur as alternating beds of very dense sand and extremely weak sandstone. The current accepted method for differentiating between soil and rock is dependent on a UCS strength occurring below 0.6 MPa. Laboratory experiments performed on samples retrieved from the Ghayathi Formation exhibit an average UCS, Point Load Index and E-modulus of 0.71 MPa, 0.11 MPa and 1128 MPa respectively. However, strength and deformation is controlled by inherent physical properties of the material such as mineral composition, density, structure, fabric and porosity. Petrographic analysis of representative samples from the Ghayathi Formation shows the sandstone comprises of highly porous fine-grains with sporadic bands of coarser material cemented together by a thin uniform crust around depositional grains with some degree of compaction. Overall the mineralogical composition by volume comprises of: calcite (80%) as a cementing agent and bioclasts, quartz (16%), rare feldspars (2%), trace amounts of pyrite (1%) and igneous rock fragments (<1%). Primary pore spaces remain unfilled and result in a high inter-particle porosity of 27.69%, which is inversely proportional to the bulk and dry densities exhibiting average values of 1968 kg/m^3 and 1660 kg/m^3 respectively. The high inter-particle porosity resulted in the degree of saturation predominantly ranging between 40–80% but the inherent water content had minimal influence on the strength of the samples. This petrographic analysis comparison to laboratory results highlights some of the complexities currently being faced in soft rocks research and further supports that improved sampling techniques, laboratory testing methods and understanding of soft rock characteristics is required for improved classification.

1 Introduction

The rapid rise and ever increasing demands placed on civil infrastructure development in Dubai and the United Arab Emirates presents a challenging prospect to all geotechnical professionals to judge their skills and abilities. One such challenge is the occurrence of soft rock underlying the quaternary aeolian sand deposits of the Ghayathi Formation. An initiative by Fugro Middle East (FME) is the continuous development

© Springer Nature Switzerland AG 2020
A. Bezvijen et al. (Eds.): GeoMEast 2019, SUCI, pp. 12–19, 2020.
https://doi.org/10.1007/978-3-030-34178-7_2

of a geotechnical database of site investigations performed though out Dubai and the region (Brouwers 2018). This paper presents a unique opportunity to investigate the Ghayathi Formation sandstone by presenting a combination of petrographic and laboratory test results in evaluating the influence physical properties such as mineral composition, density, structure, fabric and porosity has on the recorded laboratory results.

2 Geological Setting

The Arabian Gulf is a shallow elongated basin of late Pliocene to early Pleistocene age and currently has water depths rarely exceeding 100 m. The basin is asymmetric, with a gentle slope on the Arabian side and a much steeper slope on the Iranian side (Alsharhan and Kendall 2003). The basin is bound on the northwest and along the Iranian side by the Zagros mountain belt, which is the central part of the Alpine-Himalayan chain. During the late Cretaceous, a large compressive event caused the obduction of the Oman-UAE ophiolite on the eastern continental margin of the Arabian platform (Macklin et al. 2012). Eustatic fluctuations of sea level during the Quaternary, related to climate variations, resulted in shifting shorelines along the relatively flat Arabian Gulf (Al-Sayari and Zötl 2012).

A generalised overview of the Arab peninsular and stratigraphy of geology in Dubai is presented in Fig. 1. The near surface geology of coastal Dubai begins with Quaternary marine, aeolian, sabkha and fluvial deposits overlying variably cemented

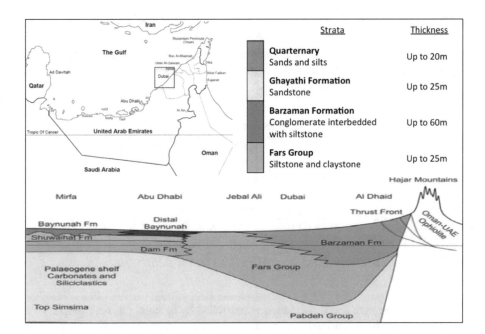

Fig. 1. Generalised overview of Dubai stratigraphy (after Macklin et al. 2012)

Pleistocene calcareous sandstone and cemented sand deposits of the Ghayathi Formation (Macklin et al. 2012). Below this, is a thick succession of fluvial sediments characterized by poorly sorted conglomerates and interbedded calcisiltites belonging to the Barzaman formation probably formed during the middle Miocene to Pliocene age (Styles et al. 2006). Underlying the Barzaman Formation is the Fars Group comprising of thinly laminated to thickly bedded gypsum-carbonate rich siltstone and claystones (Macklin et al. 2012). These rocks are currently experiencing on-going gentle deformation from the anti-clockwise movement of the Arabian plate against the Eurasion plate and localized deformation events from upwelling salt diapers.

The Ghayathi Formation was deposited during regression of sea levels and consists of a mixture of carbonate-clastic sediments (Williams and Walkden 2002). Towards the coast, it is dominantly composed of marine derived carbonates grains becoming progressively higher in quartz content inland. As the sea level retreated the sediments were reworked into a barchans dunefield where an arid climate resulted in the sediments remaining unconsolidated. Increasing wind speeds and changing wind directions later reworked the barchans dunefield into seif dunes that later became lithified before sea level transgression.

3   Results and Discussion

During data analysis of the compiled Fugro Middle East (2017) geotechnical database, eight (8) boreholes contained petrographic analysis tests encountering the Ghayathi Formation sandstone. In combination with the petrographic analysis, 26 Uniaxial Compressive Strength (UCS), 12 UCS local-strain and 41 Point Load Index (PLI) laboratory tests were performed in the same boreholes and a summary of these results is presented in Table 1. From the UCS tests, according to British Standards 5930:2015, 18.42% and 21.05% classify as extremely weak and very weak respectively. While the remaining 60.53% returned strengths less than 0.6 MPa and can be classified as very dense cemented sand.

Table 1. Summary of laboratory test results

	Moisture content (%)	Bulk density (Mg/m^3)	Dry density (Mg/m^3)	UCS (MPa)	Elastic modulus (MPa)	Poisson's ratio	Point load index (MPa)	
							I_{50}	UCS Correlated[i]
Minimum	2.30	1.70	1.38	0.03	98.00	0.14	0.01	0.05
Maximum	32.00	2.20	1.92	3.90	6271.00	0.45	0.63	2.80
Average	17.57	1.97	1.66	0.71	1128.33	0.26	0.11	0.49

NOTE: i - Correlation factor of 4,5829 (Abbs 1985; Elhakim 2015)

However, strength and deformation is controlled by inherent physical properties of the material such as mineral composition, density, structure, fabric and porosity. Therefor a 1 m selection buffer was enforced on laboratory results against petrographic analysis sample depths; where this criterion could not be satisfied the nearest laboratory results were selected and assumed as being representative. Laboratory results that satisfied these criteria are shown in Table 2.

Table 2. Representative laboratory results of petrographic analysis samples

	Petrographic sample (m)	Closest lab result (m)	UCS (MPa)	Young's modulus (MPa)	Poisson ratio	Point load index (MPa)	
						I_{50}	UCS correlated[i]
BH-A	20.55	19.04	0.80	–	–	–	
BH-B	17.45	17.20	–	–	–	0.07	0.32
		17.70	–	–	–	0.20	0.93
BH-C	28.50	28.00	0.20	–	–	–	–
BH-D	32.40	31.50	0.30	–	–	–	–
		33.10	0.70	–	–	–	–
BH-E	31.85	30.25	–	–	–	0.13	0.60
BH-F	34.10	35.00	0.06	323	0.22	–	–
BH-G	34.35	33.50	–	–	–	0.04	0.18
		33.60	0.20	–	–	–	–
BH-H	19.50	19.20	–	–	–	0.21	0.96

NOTE: i - Correlation factor of 4.5829 (Abbs 1985; Elhakim 2015)

To assess if the selected laboratory results were representative of the Ghayathi Formation sandstone: the UCS and correlated PLI were compared against all the laboratory results for the eight boreholes. Figure 2 shows this comparison: triangles represent selected representative petrographic lab results against UCS results, which are represented by hatches and correlated point load index using a determined correlation factor of UCS = 4.5829 × I_{50}, which is in general agreement with Abbs (1985) and Elhakim (2015), are represented by circles.

The laboratory results for the selected samples in near proximity of the petrographic analysis show good agreement with remaining laboratory results, as seen in Fig. 2, and are interpreted as being representative of the Ghayathi Formation sandstone.

Inspection under plane-polarised (Fig. 3a) and cross-polarised (Fig. 3b) light shows the sandstone comprises of highly porous fine-grains with sporadic bands of coarser material cemented together by a thin uniform crust around depositional grains with some degree of compaction. The remainder of the primary pore spaces remain unfilled resulting in a high inter-particle porosity. A summary of the mineralogical composition for the petrographic samples is presented in Table 3, where overall, the mineralogical composition by volume comprises of: calcite (80%) as a cementing agent and bioclasts, quartz (16%), rare feldspars (2%), and trace amounts of pyrite (1%) and igneous rock fragments (<1%).

The mineralogical composition can be used to gain a more quantitative assessment of the inter-particle porosity (ϕ) using bulk density (ρ_{bulk}) and particle density ($\rho_{density}$) as shown in Eq. 1 below.

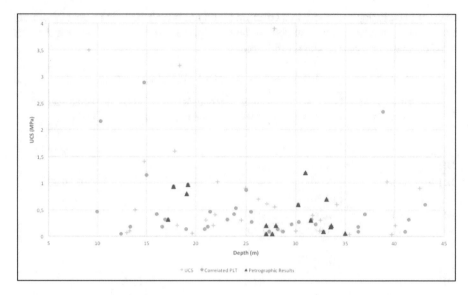

Fig. 2. Representative petrographic sample comparison

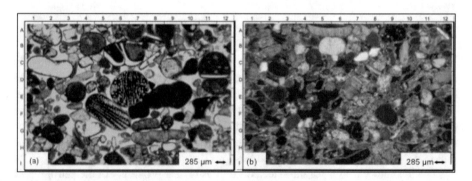

Fig. 3. Representative petrographic analysis photomicrographs: (a) Plane-polarised light showing bioclasts (D6), ooids (G2) and quartz (D8) (b) Cross-polarised light showing bioclasts (A5/6), ooids (E2), quartz (B10) and feldspar (D7)

$$\phi = 1 - \left(\rho_{\text{bulk}} / \rho_{\text{density}} \right) \qquad (1)$$

To determine the particle density: the mineral density of each of the analysed petrographic sample components where weighted and averaged by mineralogical composition by volume. From this calculation it was determined that the average inter-particle porosity for the Ghayathi Formation sandstones is 27.69%. Both the inter-particle porosity and water content decreases with increasing dry density, as shown in Fig. 4, indicating higher degrees of cementation.

Table 3. Summary of the mineralogical composition for the petrographic analysis samples

BH ID	Depth (m)	Calcite		Quartz		Feldspar		Igneous rock fragments		Pyrite		Origin
		Volume (%)	Grain size (μm)	Volume (%)	Grain size (μm)	Volume (%)	Grain size (μm)	Volume (%)	Grain size (μm)	Volume (%)	Grain size (μm)	
BH-A	20.55	75	50–400	20	20–200	2	20–100	<1	20–120			Primary
BH-B	17.45	81	50–400	16	20–200	2	20–100	<1	20–120			
BH-C	28.50	83	50–380	13	20–190	2	20–100	<1	20–120	<1	20–130	
BH-D	32.40	84	50–360	12	20–210	2	20–100	<1	20–130	1	20–120	
BH-E	31.85	78	50–330	18	20–190	2	20–100	<1	20–130	1	20–130	
BH-F	34.10	84	50–420	12	20–215	2	20–60	<1	20–120	1	20–110	
BH-G	34.35	84	50–125	12	20–150	2	20–70	<1	20–120	1	20–130	
BH-H	19.50	75	50–400	20	20–200	2	20–100	<1	20–120			

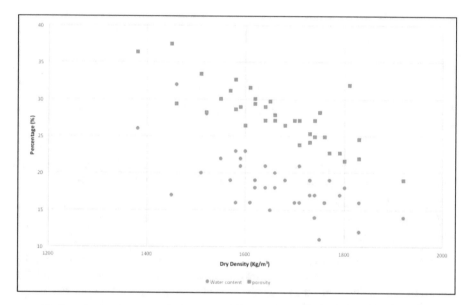

Fig. 4. Water content and inter-particle porosity for Ghayathi formation sandstone

The degree of saturation was calculated by determining the difference of saturated and unsaturated pores using the calculated inter-particle porosity and water content. Predominantly, the degree of saturation ranged between 40–80%, as shown in Fig. 5, where the degree of saturation (left axis) is plotted simultaneously with bulk density (right axis) against UCS results. The UCS shows no obvious relationship towards either the degree of saturation or bulk density, indicating that neither the moisture content nor the degree of cementation is a dominant factor influencing the strength of the rock.

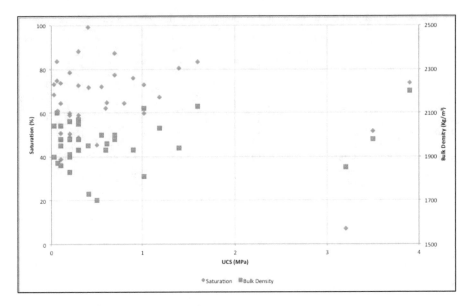

Fig. 5. Degree of saturation and bulk density vs. UCS results of the Ghayathi formation sandstone

Comparison of petrographic analysis and laboratory results highlight three main challenges currently being faced in site investigations involving soft rocks:

– Sampling techniques
– Laboratory testing
– Classification schemes.

Due to the inherent lack of strength present for these rocks, current sampling method techniques often retrieve minimal or no core recovery despite attentive effort exerted to ensure this is not the case. The lack of sample or minimal retrieved sample raises the uncertainty regarding the credibility of sampling being representative of ground conditions where the rock mass may exhibit paradoxical stable conditions during construction phase despite determined weakness. This challenge is further compounded due to the currently available and recognized laboratory testing techniques, where the results represent a bias selection of ground conditions for samples that satisfy pre determined standards and requirements to be considered appropriate for testing. Although the boundary between soil and rock is not clearly defined, as these materials occur on a continuum, the current classification schemes available to geotechnical professionals inhibit the evaluation of ground conditions, forcing the engineer to depend predominantly on arduous experience gained through trial and error dealings with such materials. Improved sampling, laboratory testing and specialised classification systems to soft rocks/hard soils will accelerate the time frame required for geotechnical professionals to confidently consider themselves competent in evaluating all the risks and hazards present at any one site.

4 Conclusion

An imminent challenge currently being faced in Dubai is the occurrence of soft rock of the Ghayathi Formation sandstone displaying average UCS, Point Load Index and E-modulus of 0.71 MPa, 0.11 MPa and 1128 MPa respectively for laboratory tests conducted on samples in near proximity of petrographic analysis samples. Review of the sandstones under increased magnification reveals the sandstone comprises a highly porous fine-grains with sporadic bands of coarser material cemented together by a thin uniform crust around depositional grains with some degree of compaction. Overall the mineralogical composition comprises of calcite (80%) as a cementing agent and bio-clasts, quartz (16%), rare feldspars (2%), and trace amounts of pyrite (1%) and igneous rock fragments (<1%). The inter-particle porosity of the Ghayathi Formation sandstone averaged at 27.69%, with the degree of saturation ranging between 40–80%. The inherent water content and bulk density of the samples showed no obvious relationship to UCS indicating neither is a dominant factor in determining strength. A combined effort in understanding and improving sampling techniques, laboratory testing and classification of soft rock/hard soils is required for the advancement of the geotechnical profession.

References

Abbs, A.F.: The use of the Point Load index in weak carbonate rocks. In Strength Testing of Marine Sediments: Laboratory and In-Situ Measurements. ASTM International (1985)

Alsharhan, A.S., Kendall, C.S.C.: Holocene coastal carbonates and evaporites of the southern Arabian Gulf and their ancient analogues. Earth Sci. Rev. **61**(3), 191–243 (2003)

Al-Sayari, S.S., Zötl, J.G. (eds.) Quaternary Period in Saudi Arabia: 1: Sedimentological, Hydrogeological, Hydrochemical, Geomorphological, and Climatological Investigations in Central and Eastern Saudi Arabia. Springer (2012)

Elhakim, A.F.: The use of point load test for Dubai weak calcareous sandstones. J. Rock Mech. Geotech. Eng. **7**(4), 452–457 (2015)

Fugro Middle East [FME]: Geotechnical Database (2017)

Macklin, S., Ellison, R., Manning, J., Farrant, A., Lorenti, L.: Engineering geological characterisation of the Barzaman Formation, with reference to coastal Dubai, UAE. Bull. Eng. Geol. Env. **71**(1), 1–19 (2012)

Styles, M., Ellison, R., Arkley, S., Crowley, Q.G., Farrant, A., Goodenough, K.M., McKervey, J., Pharaoh, T., Phillips, E., Schofield, D., Thomas, R.J.: The geology and geophysics of the United Arab Emirates: Volume 2, geology (2006)

Williams, A.H., Walkden, G.M.: Late quaternary highstand deposits of the southern Arabian Gulf: a record of sea-level and climate change. Geol. Soc. Lond. Spec. Publ. **195**(1), 371–386 (2002)

Seismic Hazard Assessment of Gujrat International Finance Tec-City

Abhishek Sharma and Sreevalsa Kolathayar[(⊠)]

School of Civil Engineering, Vellore Institute of Technology (VIT),
Vellore 632014, Tamil Nadu, India
sreevalsakolathayar@gmail.com

Abstract. The vulnerability of GIFT city from seismic hazards due to geological fault lines present in that region was assessed, and potential seismic sources were identified. GIFT city lies in Zone 3 according to seismic zonation maps of India and is nearby to the Zone 4. The past earthquake catalogue of GIFT city region is prepared. All earthquakes in the radius of 500 km from the GIFT city were taken (from United States Geological Survey), and the datasheet was prepared according to the latitude, longitude, magnitude, time and date. Spatial and temporal distribution of the foreshocks and aftershocks within 30 days from the main earthquake were analysed and declustering was done. Magnitudes of all earthquakes were homogenized and taken in terms of Moment magnitude (M_W). Seismic tectonic map of GIFT city region was taken and superimposed with these past earthquakes, and the effect of each fault line nearby was studied. Previous earthquakes were assessed, and their sources were identified. The Peak Ground Acceleration (PGA) and Ground motion attenuation were calculated using these data. The fault lines that can cause tremors or earthquakes in the GIFT city were assessed and studied.

1 Introduction

Earthquakes have a very adverse effect on human lives. They are very dynamic. Earthquakes do not give any warnings, and there is no prediction system developed for the earthquakes as of now. Though many research works are oriented to develop a prediction system for earthquake warnings, none has proved to be fruitful yet. An earthquake can prove to be catastrophic in such trend of globalization and industrialization. So the best way to deal with it is to know the seismic vulnerability of that area and to construct in that accordance. One such fast-growing state of India is Gujrat. Gujrat is undergoing very fast industrialization, and hence a lot of construction work is going on in Gujrat. One such place is Gujrat International Finance Tec-City (GIFT City). It is a part of the Delhi-Mumbai Industrial corridor that is being set up by Government of India with the assistance from Japan in this project. It is a business district being promoted by Government of Gujrat. It is India's first operational smart city in Ahmedabad. GIFT city is spread across and area of approximately 3.99 sq. Km. As two devastating earthquakes of magnitude (Ms) 8.3 and 7.7, it makes it more important to study the seismicity of the GIFT city region. The Bhuj earthquake caused mass destruction and raised questions on the building provisions. It caused the

A. Bezvijen et al. (Eds.): GeoMEast 2019, SUCI, pp. 20–28, 2020.
https://doi.org/10.1007/978-3-030-34178-7_3

breakdown of the entire infrastructure. GIFT city is the latest upcoming infrastructure project work in Gujrat. Therefore it made it mandatory to study the area and find the seismic vulnerability of the region (Figs. 1a, 1b).

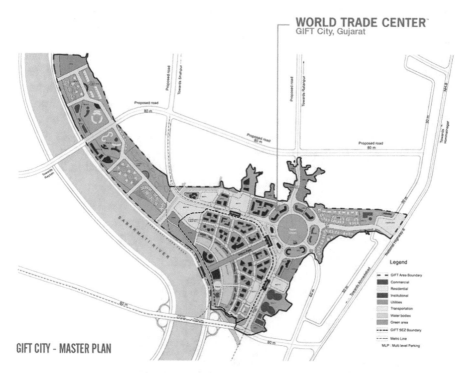

Fig. 1a. GIFT city master plan giving an overview of the city. (http://www.gujaratproperty.com/gift-city/wtc/location.html)

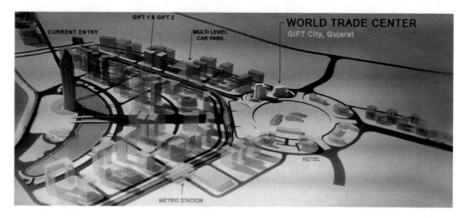

Fig. 1b. GIFT city overview (http://www.gujaratproperty.com/gift-city/wtc/location.html)

1.1 Seismicity and Tectonics of Gujrat

Gujarat has approximately 20,000 sq. Km area. The three failed rifts Kachchh, Cambay and Narmada fall in the Gujrat region. The E-W rift of Kachchh region resulted in the Kuchchh Rift basin. NE-SW Aravalli drifting trend resulted in the Cambay rift basin. The Narmada rift basin extends from the Cambay region to the southern regions of Saurashtra trend. Gujrat has very varying seismicity. A large part of Gujrat lies in Zone V and Zone 4 of seismic zonation and the rest in Zone 3. A very small north-eastern part of Gujrat lies in Zone 2. The Runn of Kachchh lies in the zone V, and a fault line passes through it. So it makes Gujrat a seismically earthquake prone area. As of March 2019, 202 earthquakes have happened in the region of 500 km from the GIFT. There is a lot of fault line in Gujrat, making it seismic vulnerable. The occurrence of the 1993 Latur earthquake followed by the destructive earthquakes of 1997 Jabalpur and 2001 Bhuj raised questions on the validity of the seismic zonation map. This further led to the revision of the seismic zonation map, and in 2002, only four zones were identified: II, III, IV and V (Fig. 2).

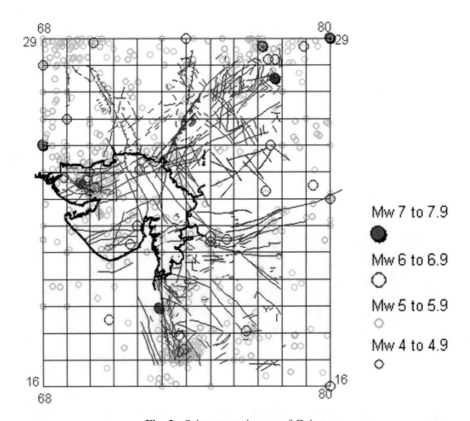

Fig. 2. Seismotectonic map of Gujrat

1.2   Earthquake Catalogue

A detailed earthquake catalogue is prepared to assess the seismic hazard. It is better to include every possible data that can be collected to get precise results for seismic vulnerability. The first step after preparing the catalogue is de-cluster it. With the main earthquake event, fore-shocks and after-shocks also take place. Even these are considered in the earthquake catalogue. Though they account for the seismic vulnerability they have lesser magnitude than that of the main earthquake. While calculating the seismic hazards, we consider the maximum magnitude earthquake, so it is necessary to de-cluster these fore-shocks and after-shocks from the earthquake catalogue as they are not the main earthquake event that took place. Moreover the data is collected from different sources so the data may repeat. So it is necessary to declutter the earthquake catalogue. The filtering of data was done based on the temporal and spatial distribution of the events. From a total of 202 events, 27 earthquake events were declustered based on their spatial and temporal distribution (Fig. 3a).

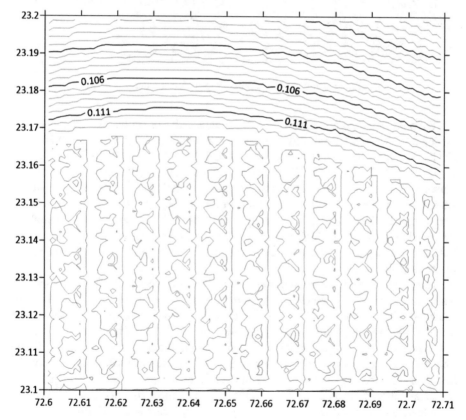

Fig. 3. Contour map showing the maximum PGA values in and near GIFT city, with respect to latitude and longitude.

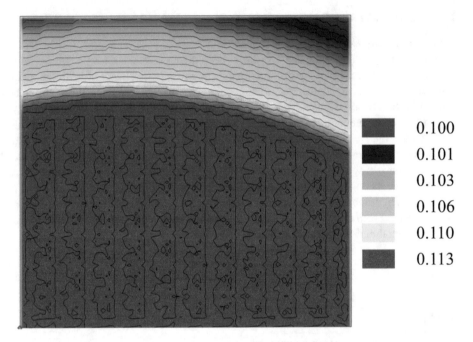

■	0.100
■	0.101
■	0.103
■	0.106
■	0.110
■	0.113

Fig. 3a. Seismic hazard contour map (colour model of Fig. 3)

The next step is to homogenize the magnitude scales of earthquake catalogue. The catalogue is prepared using old earthquake catalogues as well as data available on USGS. The historic seismic data were collected from Oldham (1883), Milne (1911), Turner (1911), Gubin (1968), Kelkar (1968), Gosavi et al. (1977) and Srivastava and Ramachandran (1983). The instrumental data were obtained from different national and international agencies like the Indian Meteorological Department (IMD), International Seismological Centre (ISC), The Incorporated Research Institutions for Seismology (IRIS), United States Geological Survey (USGS), Geological Survey of India (GSI), National Geophysical Research Institute (NGRI) and Institute of Seismological Research (ISR) (Figs. 4a, 4b).

The magnitude of the earthquakes was in different scales like Local magnitude scale, Body wave magnitude (M_B), Surface wave magnitude (M_S) and Moment magnitude (M_w). Unfortunately, the Richter scale and other magnitude scales, except Moment Magnitude, saturate after a particular intensity making it very difficult to get the exact intensity of the earthquake and making these scales unreliable. While Moment magnitude does not saturate and gives us a better idea of the intensity of the earthquake. So to compare the intensity of the earthquakes, we need to homogenize the earthquake data in terms of Moment Magnitude. The most common scale to measure the intensity of the earthquakes is the Moment Magnitude scale. The moment magnitude scale is preferred as it does not saturate. There have been many relations to homogenize the earthquake magnitude. Menon et al. (2010) developed a relation to converting the intensity scale to moment magnitude scale for earthquakes in India. The

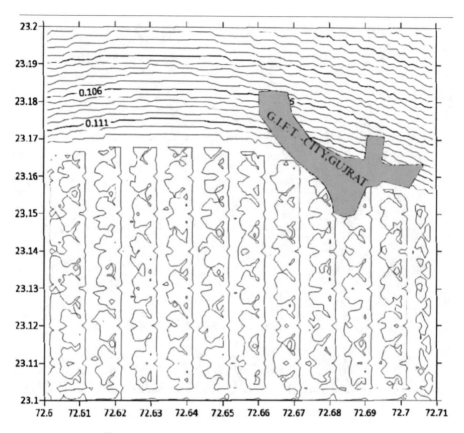

Fig. 4a. GIFT city region overlap on the contour map

surface wave (MS) and body wave (mb) magnitudes conversion relation were also given by Scordilis (2006), which were developed using the worldwide data. The magnitude conversion relations developed by Kolathayar et al. (2012) were used in this work. These relations are:

$$M_W = 1.08(\pm 0.0152)\ M_b - 0.325(\pm 0.081) \tag{1}$$

$$M_W = 0.815(\pm 0.04)\ M_L - 0.767(\pm 0.174) \tag{2}$$

$$M_W = 0.681(\pm 0.010)\ M_s - 1.993(\pm 0.053) \tag{3}$$

Now the data is homogenized and worked on in the Moment magnitude scale.

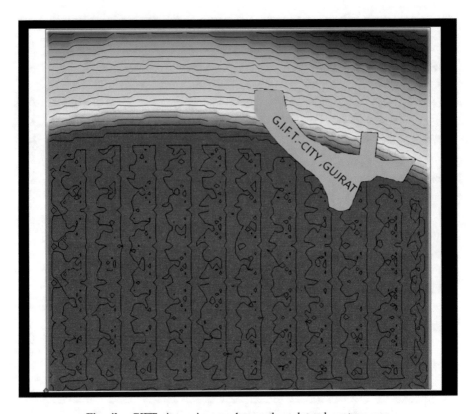

Fig. 4b. GIFT city region overlap on the coloured contour map

1.3 Deterministic Seismic Hazard Assessment

The Deterministic Seismic Hazard Assessment (DSHA) is done for a particular earthquake. DSHA uses the known data of sources to calculate the hazard. This approach uses the known sources that are in the vicinity of the site and uses the historic geological and seismic data available for the area and produces the models of ground motion at the site. It produces the ground motion models for each source and each seismic data that is taken for the study. DSHA consists of 4 steps: (a) identifying and characterizing of all sources. Identification means to identify all the potential sources in the vicinity of the site that can produce significant ground motion. Characterization means to identify source geometry and establish the potential of the earthquake that can be produced. (b) Selecting the source-site distance parameters (c) Selecting the controlling earthquake: this is the mechanism to decide the faults which will cause or have caused the earthquakes. (d) Defining the hazard for the controlling earthquake: this means to typically estimate the magnitude of the earthquake that can be produced.

The Earthquake events that have happened were superimposed on the fault lines.

1.4   Attenuation Relations

There are many ground motion attenuation relations developed by different researchers. Different attenuation relation is developed for specific areas. For the study of Gujrat region, we have used the attenuation relation by Kolathayar et al. (2012). MATLAB code was used to analyse the seismic characteristics and calculate the ground motion of the region due to the fault lines present in the vicinity. Ground motion is predicted and estimated for the GIFT city for '0' s time interval. For the analysis of the GIFT city, it was divided into the grids of 100 m × 100 m. The centre point of each square box was treated as the site, and the ground motion acceleration was calculated at that point. A total of 11211 grids were made and analysed to calculate the ground acceleration. Ground acceleration was calculated using two methods, namely, Point Source shield and Fault Shield. Attenuation relations of Raghu Kanth and Iyengar (2007), Atkinson and Boore (2006), Grazier (2016) were used for these two methods, and the maximum ground acceleration was taken. It was found out that Raghu Kanth and Iyengar (2007) gave the highest value of ground motion. Using this data, a seismic vulnerability contour was plotted for the GIFT city and the area in the immediate vicinity. This georeferenced map will provide with the seismicity of any required area in the GIFT city and will help in taking the required recommendations and building provision to construct safer infrastructure towards earthquakes.

2   Results and Discussion

The Peak ground acceleration was found out to be 0.113 g at the bedrock level. The southern region of GIFT city was found out to be more seismically active with rest to the rest. The seismic vulnerability decreases as we go towards the north-eastern zone of GIFT city. This seismic activeness was found out to be more in the southern and western region because there are more fault lines towards the southern and western parts of Gujrat. The main Tec-city lies this relatively high zone of seismicity, so they need to be much better planned with respective BIS Code provisions.

3   Conclusion

The seismically prone zones in a radius of 300 km from the GIFT city were found. The GIFT city was seismologically analysed considering the fault lines and the past earthquakes in that area. The GIFT city area into the grids of 100 m × 100 m and calculated the peak Ground acceleration at the centre of each grid using different attenuation relation suitable for the area. The contour map of the seismic vulnerability was plotted using the maximum calculated magnitudes. A map showing all the earthquakes that have happened was plotted and categorised according to their magnitudes. The seismic vulnerability of the GIFT city, Gujrat was found out. The maximum ground acceleration was found out to be in 0.113 g at the bedrock level.

References

Kolathayar, S., Sitharam, T.G., Vipin, K.S.: Deterministic seismic hazard macrozonation of India. J. Earth Syst. Sci. **121**, 1351–1364 (2012)

Atkinson, G.M., Boore, D.M.: Empirical groundmotion relations for subduction-zone earthquakes and their application to Cascadia and other regions. Bull. Seism. Soc. Am. **93**, 1703–1729 (2003)

Atkinson, G.M., Boore, D.M.: Earthquake groundmotion prediction equations for Eastern North America. Bull. Seism. Soc. Am. **96**, 2181–2205 (2006)

Boore, D.M., Atkinson, G.M.: Ground-motion prediction equations for the average horizontal component of PGA, PGV, and 5%-damped PSA at spectral periods between 0.01 s and 10.0 s. Earthq. Spectra **24**, 99–138 (2008)

Kolathayar, S., Sitharam, T.G., Vipin, K.S.: Spatial variation of seismicity parameters across India and adjoining areas. Nat. Hazards (2011). https://doi.org/10.1007/s11069-011-9898-1

Vipin, K.S., Anbazhagan, P., Sitharam, T.G.: Estimation of peak ground acceleration and spectral acceleration for south India with local site effects: Probabilistic approach. Nat. Hazards Earth Syst. Sci. **9**, 865 (2009)

Vipin, K.S., Sitharam, T.G.: Multiple source and attenuation relationships for evaluation of deterministic seismic hazard: logic tree approach considering local site effects (2010). https://doi.org/10.1080/17499518.2010.532015

Raghu Kanth, S.T.G., Iyengar, R.N.: Estimation of seismic spectral acceleration in peninsular India. J. Earth Syst. Sci. **116**, 199–214 (2007)

Vipin, K.S., Sitharam, T.G., Kolathayar, S.: Assessment of seismic hazard and liquefaction potential of Gujarat based on probabilistic approaches. Nat. Hazards **65**, 1179–1195 (2013)

Anbazhagan, P., Vinod, J.S., Sitharam, T.G.: Probabilistic seismic hazard analysis for Bangalore. Nat. Hazards **48**, 145–166 (2009)

Campbell, K.W., Bozorgnia, Y.: Updated near-source ground motion (attenuation) relations for the horizontal and vertical components of peak ground acceleration and acceleration response spectra. Bull. Seism. Soc. Am. **93**, 314–331 (2003)

Cramer, C.H., Kumar, A.: 2001 Bhuj, India, earthquake engineering seismoscope recordings and Eastern North America ground motion attenuation relations. Bull. Seism. Soc. Am. **93**, 1390–1394 (2003)

Gutenberg, B., Richter, C.F.: Seismicity of the Earth and Associated Phenomena, 2nd edn. Princeton University Press, Princeton (1954)

Mandal, P., Kumar, N., Satyamurthy, C., Raju, I.P.: Ground-motion attenuation relation from strongmotion records of the 2001 Mw 7.7 Bhuj earthquake sequence (2001–2006), Gujarat. India. Pure Appl. Geophys. **166**, 451–469 (2009)

Menon, A., Ornthammarath, T., Corigliano, M., Lai, C.G.: Probabilistic seismic hazard macrozonation of Tamil Nadu in Southern India. Bull. Seism. Soc. Am. **100**, 1320–1341 (2010)

Characterising and Structural Review of the Rock Mass and Its Geological Structures at Open Pit Mine in Queensland-Australia

Maged Al Mandalawi[1]([✉]), Manar Sabry[1], Greg You[1],
and Mohannad Sabry[2]

[1] Faculty of Science and Technology, Federation University,
Mt. Helen 3350, VIC, Australia
sp.group@ymail.com
[2] School of Computing, Engineering and Mathematics,
Western Sydney University, Sydney, Australia

Abstract. Rock characterisation is important to the feasibility of the Handlebar Hill open cut mine at Mt Isa, Queensland-Australia, because of the complex structural geology and the diversity of slope formations. The different rocks are affected by complete and moderate oxidisations coupled with mining works, giving rise to potential slope instability. Through the characterisation of these rocks, there is more confidence in the prediction of their behaviors in terms of failure mechanisms and slope stability. The objective of this research was to evaluate the properties of the pit rock masses. The geotechnical engineering practice approach was based on defining the parameters of the Hoek-Brown, Barton-Bandis and Mohr-Coulomb failure criteria. A program involving investigation that included field measurements, laboratory tests, hydrogeological settings, empirical indices and findings using the RocLab program was applied. The inputs help to analysis of pit slope stability and to understand the effects of different pit configurations on slope performance to allow safe and economic mining operations.

Keywords: Geological structures · Orebody deposits · Lithological domains · Intact rock

1 Introduction

Handlebar Hill mine operation is located in Mt. Isa nearly 1,850 km NW of Brisbane, Queensland, Australia, as shown in Fig. 1a. The mine was open pit mining to extract zinc and lead deposits. The mining formally started in 2008. To date, the pit had two stages of excavation with a further third stage involving expansion to the west planned following improvements in mining methods and metallurgical performance (Fig. 1b). This third stage of development involves a narrow cutback to the south of the pit that will enable deeper ore from under the first two stages of development to be targeted. This will add approximately 1 million tons of ore containing 8.7% zinc, 2.7% lead and 44 g/t silver to the Handlebar Hill open cut pit reserves (Glencore Coal 2012).

© Springer Nature Switzerland AG 2020
A. Bezvijen et al. (Eds.): GeoMEast 2019, SUCI, pp. 29–51, 2020.
https://doi.org/10.1007/978-3-030-34178-7_4

The Handlebar Hill open pit is approximately 1 km in length, 0.5 km in width, and between 170 m and 180 m in depth. The two cutbacks of the pit and the final pit shell are shown in Fig. 2. The north wall has a maximum overall slope angle of 40°, the south wall is 45°, the east wall is 40° and the west wall is approximately 45°. The mine pit slopes have been very stable except at the west wall, which has experienced a few local bench scale failures.

This paper describes the effects of different Handlebar Hill open pit configurations on the slope performance as it relates to geotechnical, hydrological, geometrical and many other environmental and engineering conditions. Furthermore, the influence of slope stability on the economy and safety of mining the west wall of the pit are assessed, and improvement opportunities are identified through detailed geotechnical investigation and implementation.

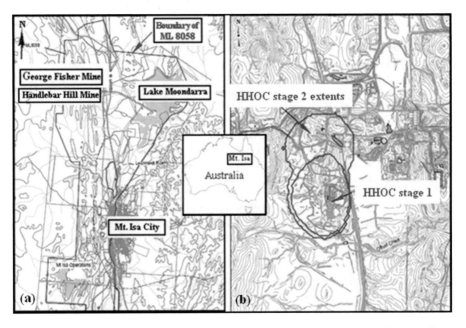

Fig. 1. (a) Location map of the Handlebar Hill mine in Mt Isa and (b) extent of the Handlebar Hill open pit for stages 1 and 2 (after AGE 2007)

Fig. 2. The current Handlebar Hill open pit with two cutbacks, looking south

1.1 Orbody Resources

The location, orientation and geometry of orebodies play a major role in the pit slope design and the process of ore extraction. According to Neudert (1983), in the vicinity of the economic deposits (Mt Isa-copper, silver, lead and zinc; Handlebar Hill, Hilton and George Fisher-silver, lead and zinc), the mineralisation is predominantly located within the Urquhart Shale, which tends to dip southwards and westwards with susceptible stratigraphy.

A cross-section of the orebody corridors within the Handlebar Hill open pit at 4,760 N is illustrated in Fig. 3. This image shows the zinc orebody (orange) with another five orebodies that comprise the mineral resources. The faults geometries and leached zones are identified by dashed red lines (after Rosengren 2007).

The Handlebar Hill orebody corridors are elongated and dip sub-vertically to the west of the pit. These orebodies extend close to the surface and to a depth of more than 500 m. The final pit has reached a depth of 180 m (Fig. 3). Since the Handlebar Hill open pit operation was opened, mining rates have been between 40 and 80 mtpa, with ore production of 17 mtpa at 0.77 g/t gold and 0.18% copper. The ore reserve was 120 mt at 0.79 g/t gold and 0.17% copper (Glencore Coal 2012).

The characteristics of the orebodies of the Handlebar Hill open pit are largely determined by the geological conditions of the rock masses. Therefore, it would be advisable to investigate the geological context of the rocks surrounding the pit excavation.

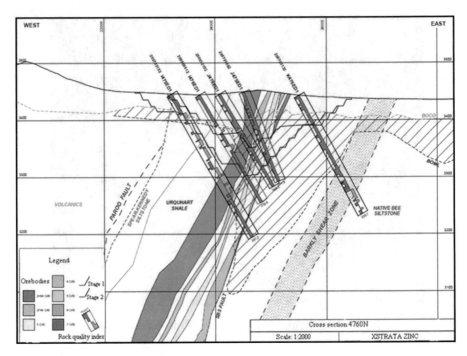

Fig. 3. Cross-section at 4,760 N showing the typical lithology, orebody deposits and major geological structures of the Handlebar Hill final open pit shell (after Rosengren 2007)

2   Geological Contexts

Geological and geotechnical data relating to the Handlebar Hill open pit was collected between 2006 and 2009. This data was verified and used over the past six years for mine tasks ranging from the project evaluation to the present production.

The Handlebar Hill open pit is an up dip extension of the George Fisher deposit stereography. Core logging operations encountered several rock domains in the Mt Isa group. There are six distinct groups of rocks including Eastern Creek Volcanics, Magazine Shale, Kennedy- Spears Siltstone, Urquhart Shale and Native Bee Siltstone. Figure 4 illustrates the general surface geology of the Handlebar Hill open pit.

Chemical weathering of the Handlebar Hill open pit rocks is generally shallow: the Eastern Creek Volcanics are extremely weathered, the Magazine Shale is slightly weathered, and the largest parts of the Spears Siltstone and Uquhart Shale are fresh. Oxidation has altered the zone that extends from the surface to the base of complete oxidation (BOCO), which is defined in Fig. 3 by a dashed orange line. BOCO is where the rock is still largely intact but is decomposed and may disintegrate. This zone is variable in depth, with a minimum of approximately 30 m over the Kennedy-Spear Siltstone and a maximum of approximately 120 m over portions of the mineralised Urquhart Shale. Oxidation and leaching affect only the uppermost 50 m of the west wall, but weak expressions are present at shallower levels that are associated with small to medium structures and contact between geological units.

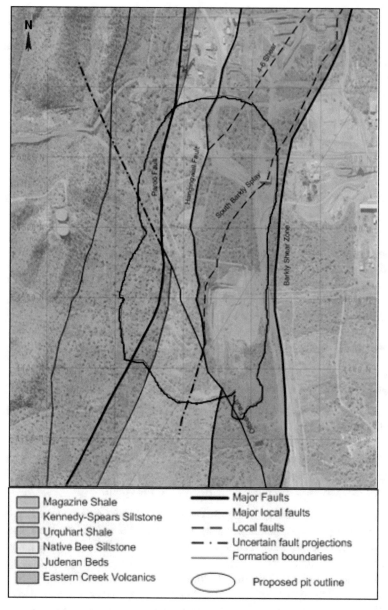

Fig. 4. Lithological domains intersecting Handlebar Hill open pit (after HH feasibility study 2007)

Rocks that are deeper than the level of BOCO have been identified to be moderately leached within the footwall of the Urquhart Shale, which is defined in Fig. 3 by a dashed purple line. The base of moderate leaching (BOML) is where 50% or less of the rock is decomposed and may disintegrate into soil, but is still largely fresh and has not changed colour. At the east and in the middle of the pit, this zone of leaching extends to the final floor of the pit.

Valenta (1988) described the Mt. Isa rock group in terms of the formation of mineralisation and the major geological structures of the region. He recognised multiple stages of rock displacement and described major changes that have occurred in compression and extension of the rock masses in the region. Therefore, different groups of major faults developed at Handlebar Hill during the regional deformation processes. Faulting is the primary geological structure within the Handlebar Hill open pit mine and is classified as followings.

Strike Slip Faults

Three large-scale geological structures can be classified as strike slip faults, and each is separated by a major transverse fault. The entire mining area is enveloped by major strike slip faults. The major strike slip faults that have been recognised in the Handlebar Hill open pit are as follows:

- The Paroo fault is a commonly highly deformed zone of substantial shearing fabrics and is up to 200 m wide. The fault is of N-S orientation and dips within 70° to the west of the deposit.
- The Barkly Shear zone is highly sheared zone of 20–60 m wide. The zone is of N-S orientation and dips within 60°–70° to the west.
- The hangingwall fault is a combination of several separated faults. The orientation is variable but is mostly to the N-S and dips to the west within 60° to sub-vertical.

Transverse Faults

The major deposit scale transverse faults that have been identified as follows:

- The 4–6 Shear is a sheared weak zone of 10 m wide that is orientated to NE-SW and dips within 60°–70° to the NW.
- The Offset fault is a mostly sheared zone that dips steeply towards NW-SE within 70° south of the deposit.
- The South Barkly Splay fault is a sheared weak zone of 10 m wide that is orientated to NW-SE with a steep dip within 70°–75° to the south.

The details of pit discontinuities have been obtained by outcrop and geological surface mapping surveys, laboratory tests of rock specimens, and analysis of diamond core logging samples of weathered and fractured rocks, and assessing the nature of the surfaces of the geological structures. A description of the general geological structures is shown in the two-dimensional architecture of the pit surface in Fig. 5. Based on these large-scale geological structures, slope domains were described and used for the evaluation of the geotechnical design domains and are discussed in the next section.

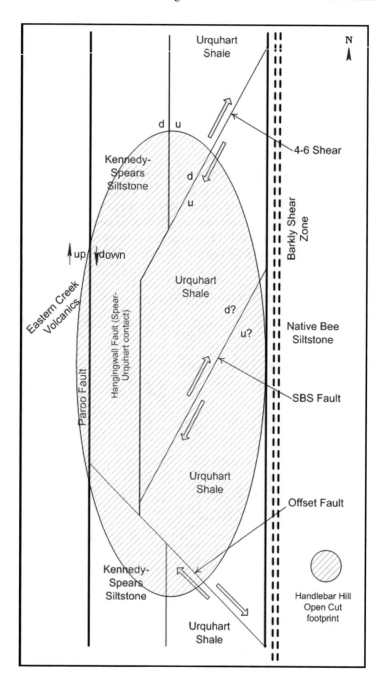

Fig. 5. Schematic description of the general geological structures showing the two-dimensional architecture of major geological structures of the Handlebar Hill open pit (after Valenta 1988)

3 Geotechnical Characterisation

Characterisation of pit rock involves identifying the properties of the current state of intact rock and geological structures. Characterisation of pit rock encompasses rock mechanics and rock engineering as well as describing the standard procedures that are used to obtain physical and mechanical strength properties following core.

3.1 Intact Rock Properties

Over 4,000 specimens were tested from the zone around Handlebar Hill. Laboratory tests included the intact rock's uniaxial compressive strength (UCS) and the dynamic modulus, density and tensile strength for both weathered and fresh rocks. Tables 1 and 2 summarise the UCS and dynamic modulus for lithological units in the area of study. In the UCS tests, failures were controlled by weak planes, the cohesion value ranged from 0 to 20 MPa, and the frictional angle ranged from 6° to 65° (Seville 1981).

Table 1. UCS and dynamic modulus data of oxidised and leached rock (after Seville 1981)

Domain	Number of samples	UCS (MPa)				Dynamics modulus (GPa)			
		Mean value	Standard deviation	Max. value	Min. value	Mean value	Standard deviation	Max. value	Min. value
Eastern Creek Volcanics	101	36.4	21.2	102.0	4.3	48.4	19.6	97.7	15.3
Spears Siltstone	18	26.3	15.5	77.6	7.7	30.4	16.1	58.7	11.0
Urquhart Shale	186	32.7	32.5	207.6	1.9	41.0	24.6	165.8	7.3

Table 2. UCS and dynamic modulus data of fresh rock at IMB (after Seville 1981)

Domain	Number of samples	UCS (MPa)				Dynamic modulus (GPa)			
		Mean value	Standard deviation	Max. value	Min. value	Mean value	Standard deviation	Max. value	Min. value
Judenans	34	123.9	58.8	327.8	45.1	71.5	13.6	99.4	40.6
Eastern Creek Volcanics	320	55.5	35.9	211.5	4.6	63.9	21.9	110.6	16.4
Magazine Shale	10	31.8	17.5	58.0	5.3	50.7	10.1	66.9	38.3
Spear Siltstone	709	111.1	81.9	401.8	2.8	75.4	11.9	135.1	12.1
Urquhart Shale	1102	107.9	78.8	450.9	3.1	87.5	21.0	230.3	14.1
Native Bee Siltstone	301	131.2	66.4	305.5	6.3	80.2	9.3	116.3	38.5
Breakaway Shale	73	125.9	81.9	333.0	8.5	58.7	10.7	79.9	37.1

3.2   Methods of Characterisation

The determination of the generalised Hoek-Brown strength parameters of a rock mass (m_b, s and a) is based on obtaining, the UCS parameter of intact rock, the rock mass deformation modulus (E_{rm}), the intact rock parameter (m_i), the rock mass geological strength index (GSI) and the blasting disturbance factor (D).

Statistically analysis of the results obtained from tests of UCS was used to derive the generalised Hoek-Brown intact uniaxial compressive parameter (σ_{ci}) with the empirical material value of Hoek-Brown constant (m_i), as shown in Table 3.

The rock mass deformation modulus (E_{rm}) requires the intact rock deformation modulus (E_i), the blasting disturbance factor (D) and the rock mass geological strength index (GSI) as input data. The simplified Hoek and Diederichs (2006) is then used to obtain the value of the rock mass modulus of deformation (E_{rm}).

In practice, the qualities of rock masses in the Handlebar Hill and George Fisher areas were classified according to Bieniawski's rock mass rating system. The values of the GSI were directly estimated from (Bieniawski 1989), after setting the water table and the adjustment of joint orientations and ranged from 32 to 20 for poor rocks.

The advantage of the generalised Hoek-Brown failure criterion in the RocLab (Rocscience 2014) software is that the estimates of rock mass strength parameters appear to be realistic (Simmons and Simpson 2006). The equivalent Mohr-Coulomb strength envelope was obtained by implementing the same approach. For a specific generalised Hoek-Brown failure envelope, through this approach, Mohr-Coulomb envelope equivalent to the Hoek-Brown envelope over the slope stress ranges was calculated. Hoek et al (2002) conducted a related slope analysis study, using Bishop's circular failure method for including a variety of rock mass properties and slope geometries, as given in Eq. (1):

$$\sigma'_{3max} = 0.72 * \sigma'_{cm} \left(\frac{\sigma'_{cm}}{\gamma H} \right)^{-0.91} \tag{1}$$

where σ'_{3max} is the major effective principal stress at slope failure (which presents the equivalent of Mohr-Coulomb strength to the generalised Hoek-Brown envelope over the slope stresses), (H) is the height of the slope, (γ) is the unit weight of the slope rock mass, and (σ'_{cm}) is the rock mass strength, which can be calculated from Eq. (2):

$$\sigma'_{cm} = \sigma_{ci} \frac{(m_b + 4s - a(m_b - 8s))(\frac{m_b}{4} + s)^{a-1}}{2(1+a)(2+a)} \tag{2}$$

Where m_b is the reduced value of m_i, which is the intact rock material constant, and s and a are rock mass constants. The shear strength parameters of different rocks at the pit slopes were described using this method. To calculate the cohesion and the internal friction angle of the Mohr-Coulomb equivalent to the generalised Hoek-Brown envelope of the fresh Eastern Creek Volcanics rock, the space between the linear model and the Hoek-Brown curve was minimised. This involves the geometrical fitting of the Mohr-Coulomb envelope in which the area above the line of the Mohr-Coulomb envelope is equal to the area below the line as illustrated in Fig. 6, for fresh Eastern

Creek Volcanics rock with unit weight 28.3 KN/m^3, height of 180 m, $\sigma'_{3max} = 3.5$ MPa, $\sigma_{ci} = 55$ MPa, GSI = 43, $m_i = 4.07$ and D = 0.7. The Mohr-Coulomb strength parameters can be estimated for the same model and are presented in Fig. 6.

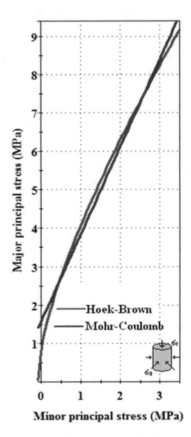

Fig. 6. Principal stresses plot of Mohr-Coulomb and Hoek-Brown criteria for fresh Eastern Creek Volcanics rock

For the slope stability analysis, UCS test results showed a significant decrease in rock strength when using samples of larger sizes (Hoek and Brown 1997). Sjoberg (1997) found that during an investigation of rock mass strength at the Aznalcollar open cut mine in Spain, there was good agreement between the estimated GSI value and the strength value calculated from back analysis of rock failures in the footwall of the pit. Caia et al. (2007) suggested that the reduction in the estimated GSI is reasonable. Therefore, for the slopes of the Handlebar Hill open pit, it is important to emphasise the variation in strength estimates obtained from tests of different disturbed rock samples. It is expected that different locations on the slope have endured different degrees of weathering, stresses re-distribution and discontinuities in orientation. This uncontrolled problem resulted in unrealistic strength values being obtained from laboratory work. Eventually, the back analysis of a bench rock failure at the west slope of the Handlebar

Hill open pit, showed shear strength values that were lower than those obtained from laboratory-tested rock samples.

A series of studies based on the UCS test have been performed on Hilton mine rocks. Results of many tests of samples of Urquhart Shale showed that the UCS value decreased with an increase in sample diameters (Review of Hilton Mine 1981). Medhurst and Brown (1996) tested four different sizes of coal samples from the Moura mine in Queensland and the triaxial compression testing showed different shear strength results for the rock specimens of different diameters.

For further geotechnical information about Mt Isa and Hilton mine rock characterisations, a variety of samples have been subjected to the laboratory direct shear test. The mechanical behaviours of the samples showed different values for shear and normal stresses for the same rock. Typical results of the application of direct shear tests to Urquhart Shale rock specimens from the Hilton mine are shown in Fig. 7.

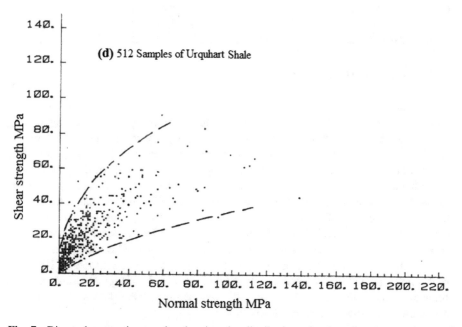

Fig. 7. Direct shear testing results showing the distribution of values for shear and normal stresses obtained for 512 samples of Urquhart Shale from the Hilton mine at Mt Isa (after review of Hilton Mine 1981)

3.3   Results and Conclusion

The rock mass properties are summarised in Table 3. The rock mass Hoek - Brown parameter, m_b, ranges from 0.12 to 0.52, s is nearly 0.0 and a ranges from 0.508 to 0.515. The cohesion varies from 0.5 to 4.2 MPa and the frictional angle ranges from 12.11° to 14.37°. The rock mass modulus of elasticity, E_m, ranges from 1,398 to 5,577 MPa, tensile strength ranges from 29 to 305 kPa and compressive strength ranges from 279 to 2191 kPa.

The Handlebar Hill open pit rock strengths range from high (UCS > 131 MPa) to low (UCS < 26 MPa). Rock mass shear strength is typically determined by RocLab (Rocscience 2014), and was used to characterise the rock mass strength in this paper.

Table 3. Rock mass strength parameters

Rock Formation			Green stone oxidated	Spear Siltstone oxidated	Urquhart Shale oxidated	Green stone	Magazine Shale	Spears Siltstone	Urquhart Shale	Native Bee Siltstone
Hoek – Brown Criterion	m_b	$\times 10^{-5}$	0.122	0.145	0.120	0.178	0.175	0.116	0.470	0.520
	s		9.59	15.8	9.59	26.1	26.1	31.9	35.3	31.9
	a		0.515	0.512	0.515	0.509	0.509	0.508	0.508	0.508
Mohr – Coulomb Fit	c	MPa	0.628	0.499	0.560	1.170	0.666	2.007	3.311	4.166
	ϕ	°	12.22	13.19	12.13	14.37	14.27	12.11	20.58	21.31
Rock mass parameters	σ_t	MPa	−0.029	−0.029	−0.026	−0.081	−0.047	−0.305	−0.081	−0.081
	σ_c	MPa	0.311	0.299	0.279	0.832	0.477	1.856	1.904	2.191
	σ_{cm}	MPa	1.556	1.259	1.386	3.015	1.713	4.968	9.559	12.192
	E_{rm}	MPa	1,852	1,398	1,568	3,589	2,848	4,604	5,577	4,897

4   Geological Structures Properties

The mechanical properties of the major and minor geological structures of the rocks from Mt Isa region were obtained from numerous tests of rock cores by mining researchers over more than 30 years. More recently, most of laboratory work at Handlebar Hill open pit has focused on testing the properties of intact rocks. The properties of different structures used in the slope stability analysis of the pit are based on both back-analysis and geological analysis of diamond core logging samples. Oxidised and moderately weathered specimens from all domains that can alter the mechanical properties of the rock were tested and specifically measured. The borehole locations for geotechnical diamond core drilling at the Handlebar Hill open pit are shown in Fig. 8, with a legend listing the borehole site code, date and depth.

The shear strength of the Handlebar Hill open pit's geological structures is mostly based on test results summarised by Tarrant and Lee (1981) of typical Urquhart Shale formation discontinuities and in particular low strength bedding planes; they reported the following classes of geological structures:

- For very rough joints and dolomite veins, the maximum cohesion (c) is 31 MPa and the maximum friction angle (φ) is 51°.
- For smooth graphitic bedding planes, the minimum $c = 2$ MPa and the minimum $\varphi = 12°$.

Unconfined compression tests have been used to assess the shear and normal stresses that failed on a fracture that corresponded with the triaxial strength of the rock mass and the shear strength of the weakest fractures in the specimen. The data analysed by Tarrant and Lee (1981) was collected over 50 years for typical Mt. Isa regional rocks.

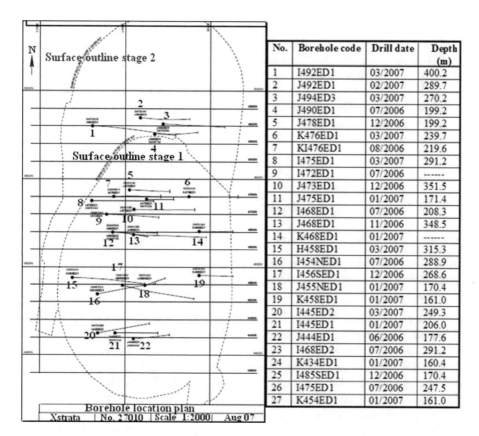

No.	Borehole code	Drill date	Depth (m)
1	I492ED1	03/2007	400.2
2	J492ED1	02/2007	289.7
3	J494ED3	03/2007	270.2
4	J490ED1	07/2006	199.2
5	J478ED1	12/2006	199.2
6	K476ED1	03/2007	239.7
7	KI476ED1	08/2006	219.6
8	I475ED1	03/2007	291.2
9	I472ED1	07/2006	------
10	J473ED1	12/2006	351.5
11	J475ED1	01/2007	171.4
12	I468ED1	07/2006	208.3
13	J468ED1	11/2006	348.5
14	K468ED1	01/2007	------
15	H458ED1	03/2007	315.3
16	I454NED1	07/2006	288.9
17	I456SED1	12/2006	268.6
18	J455NED1	01/2007	170.4
19	K458ED1	01/2007	161.0
20	I445ED2	03/2007	249.3
21	I445ED1	01/2007	206.0
22	J444ED1	06/2006	177.6
23	I468ED2	07/2006	291.2
24	K434ED1	01/2007	160.4
25	I485SED1	12/2006	170.4
26	I475ED1	07/2006	247.5
27	K454ED1	01/2007	161.0

Fig. 8. Borehole locations for geotechnical core drilling undertaken at the Handlebar Hill open pit with a legend listing the site code, date of drillings and depth of each borehole (after Rosengren 2007)

In slope stability analyses of Handlebar Hill open pit, the cohesion of structures ranged from 6 to 18 MPa and the friction angle ranged from 13° to 25°. From these analyses, the behaviours of the shear strength of the geological structures can be presented using Barton-Bandis and Mohr-Coulomb failure criteria.

The presence of groundwater can have a major effect on the stability of a pit slope due to the groundwater's influence on rock's mechanical parameters, which are derived from core logging and measured in the laboratory. The definition of the hydrological characterisation of the pit is, therefore, an important step in this investigation.

5   Hydrological Condition and Impacts

It has long been recognised that rainfall and groundwater levels critically influence the stability of slopes in several ways. Rainfall at the Handlebar Hill open pit can occur at any month during wet season, but mainly occurs during the wet season from November to March. Average monthly rainfall and pan evaporation data for the period from 1932 to 2004 is summarised in Table 4 (AGE 2007). Evaporation significantly exceeds rainfall in the area, and there is more than 90 mm of rainfall in both January and February every year. This seasonal rainfall has trigged slope instability and preceded some of the localised bench failure in the west slope. Therefore, during production in 2011, the Handlebar Hill open pit site was shut down for 12 h after every rainfall event of 15 mm or more in one day.

Table 4. Climate data for Mt Isa mines (after AGE 2007)

	Jan	Feb	Mar	Apr	May	Jun	Jul	Aug	Sep	Oct	Nov	Dec	Mean Annual
Average monthly rainfall (mm)	91	91	60.7	15.3	17.2	10.3	6.7	3.4	7.1	19.9	28.6	60.2	411.4
Average daily pan evaporation (mm)	9.9	9.2	9.1	8.1	6.2	5.1	5.3	6.6	8.6	10.3	11.2	11.2	8.4
Average monthly pan evaporation (mm)	306.9	259.9	282.1	243	192.2	153	164.3	204.6	258	319.3	336	347.2	3,066.5

Another factor that can influence the stability of the west slope of the pit is the increasing of infiltration of rainfall into the rock masses. Along the natural hill adjacent to the west wall, water can flow onto the west slope as a result of direct precipitation run on from the top of the hill. The water infiltration increases the pore pressure, softens rock and may cause erosion to rock slopes (Moshab 1999). In 2010, this direct precipitation was controlled and surface flow was diverted from the adjacent hill through two channels before reaching the pit wall.

At the west wall, groundwater pressure behind the slope was highlighted when seeping water was visible between RL3,400 and RL3,380. Rosengren (2008) suggested a design for a depressurisation system to be installed along the toe of the RL3,400 and RL3,380 face. Another extension of drainage through the major Paroo Fault into the Eastern Creek Volcanics was suggested to determine whether water inflow is increasing in this rock mass.

Oxidation and leaching limits within the Handlebar Hill open pit, specifically in the Magazine Shale, Spears Siltstone and Urquhart Shale, have formed permeable areas within these rocks (Fig. 9). Mapping of the base of complete oxidation and the base of moderate leaching are integral to the geological model that forms a major component of the assessment of the extent of aquifer units (AGE 2007). The most important factor to be considered for slope stability analysis is the position of the base of complete oxidisation, which is within the pit at some irregular depth. The base of complete oxidisation influences the mineralogy and rock mass properties (McKnight 2015) and has been mapped because of its significance in several areas, including processing and possibly pit design.

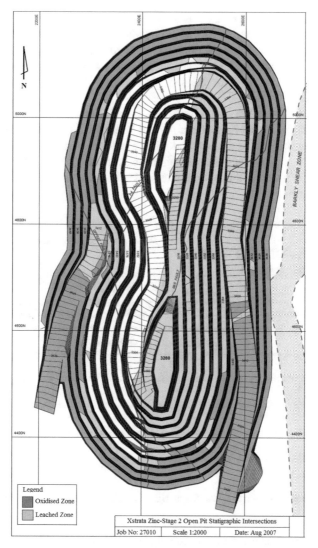

Fig. 9. Oxidised and leached zones in the pit stratigraphic intersections map (after Rosengren 2008)

Previous geological studies conducted by Mt Isa mines identified three discrete water aquifers defined by the contact between different geological domains. The unconfined aquifers are the ore-zone, which is the main aquifer; the footwall aquifer, which is contained within the leached zones of the Native Bee Siltstone; and the hangingwall aquifer, which is situated in the quartzite of dominated Judenan beds (AGE 2007).

A manual analytical valuation was undertaken to estimate the water level in the pit and the probability of water overflowing or groundwater discharge (AGE 2007). The initial groundwater sources are the rainfall and the run-off accumulation. Therefore, the pit water table level will be higher than in the aquifers. This will result in water flow from surrounding areas towards the pit. Over a long period, the pit may turn to a groundwater sink.

The result of this analysis is presented in Fig. 10. The maximum recharge scenario of high groundwater inflow is at an equilibrium level of 3,326 m RL. This is equivalent to 124 m below the ground surface and will be reached in more than 150 years. The minimum recharge scenario and an equilibrium level of 3,296 m RL is equivalent to 154 m below the ground surface and may be reached in more than 200 years.

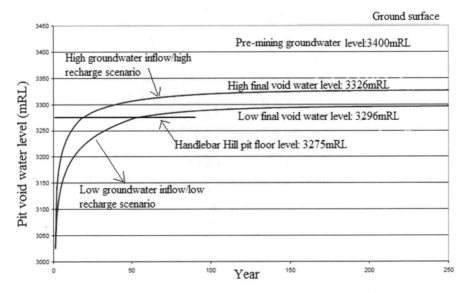

Fig. 10. Recovery of aquifers of Handlebar Hill open pit and George Fisher mines (after AGE 2007)

The pre-mining water tables indicated that there is an acceptable capacity for the pit system to absorb a substantial volume of water with no risk of over-flowing or groundwater discharge. During the next decades, increased evaporation from the pit will result in a reasonable reduction in the pit groundwater level to the equilibrium levels shown in Fig. 10.

The Handlebar Hill open pit hydrogeological conditions are complex and are controlled by large-scale structures, as explained in the geotechnical section. Generally, dry conditions in winter increase pit water evaporation and reduce the groundwater level. The average groundwater inflow in the area north of the pit is around 17.4L/s. In many events, and particularly after the circular-toppling failure that occurred on the west slope 80 m below the ground surface at RL3368, the inflow of water was visible on the slope's surface. Most of the rock failure events were localised and followed heavy rainfall.

6 Pitt Wall Blasting

In many slope excavations, pre-splitting blasting was conducted in fresh rock masses or as a blasting course prior to trim blasting. The Spear Siltstone at the west slope is basically fresh and pre-splitting was first carried out followed by trim blasting on the face below.

Trim blasting involves leaving 20–25 m of material against the final wall and blasting this to a free face using smaller blast hole diameters and reduced powder factors (Rosengren 2008). Generally, trim blasting works well on the weak or highly fractured rock masses and results in smooth blasting.

7 Pit Slope Monitoring

7.1 Visual Monitoring

Because open pit slope failures often occur after rain, visual monitoring directed by geotechnical staff is a common method of evaluating pit slope stability after rainfall. Most failures occur where there is prior evidence of cracking, mostly on the berms. At the Handlebar Hill open pit, each of the accessible berms in the pit is walked on after every rainfall to be checked for evidence of cracking. The inspection was recommended to be conducted on a weekly basis, even when there is no rain. The geotechnical engineer marks each visible deformation crack using red spray paint, but monitoring pin installation on either side of the crack is preferred. A verbal notice is then directed to the staff and operators of equipment around the area to bring the risk of the cracks to their attention. In cases where major cracking develops, or a large area of the slope is affected, more sophisticated monitoring such as automated crack-meters or extensometers needed to measure the rock deformation. Up to now, none of automated methods have been used for in situ measurements. SiroVision software has been used to obtain three dimensional digital photographs and is broadly used on slopes to identify high-risk areas and apply the correct visual monitoring systems.

The Ground-Probe Slope Stability Radar
Ground-probe slope stability radar was installed at the east monitoring station of the Handlebar Hill open pit in June 2008 to monitor rock movements on the west wall. The west monitoring station is operated by laser scanning against the stable east slopes. The locations of the two base stations and monitoring prisms are shown in Fig. 11. Results of radar monitoring are telemetered to a central computer in the mine geologist's office.

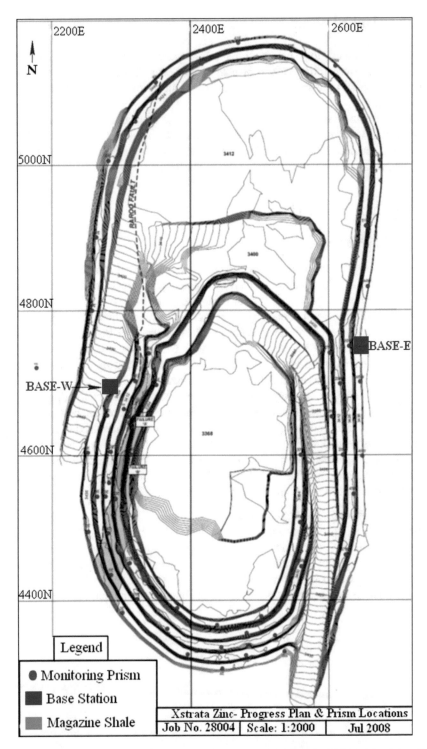

Fig. 11. Prism network locations and monitoring set-up fitted in July 2008 at Handlebar Hill open cut pit (after Rosengren 2008)

8 Handlebar Hill Open Pit Slope Design and Implications of Mining

The Handlebar Hill open pit was designed for two stages of excavations with progressive cut-backs and a maximum depth of 170–180 m. Due to the geometrical, geological and geotechnical conditions at the site, the west slope of the pit was divided into three sections, each with its own geotechnical characteristics (Read 2001). These characteristics control the stability of the pit depending on the orientation of structures in terms of slopes and different geotechnical domains.

The geotechnical design of the pit slopes is based on local lithology and estimation of the probable modes of failure, including those controlled by, rock mass properties and geological structures. Geological structures are normally the controlling factor, and in the case of fresh rocks, the slope orientation may influence the design criteria. As such, the structures in a particular geotechnical domain may have a greater potential for structurally controlled instability when combined with a particular slope orientation. For a different slope orientation in the same domain, the potential for structurally controlled instability may be different. Therefore, based on results of kinematic analyses, a further subdivision of the west slope into sectors may be required. The historical performance of pit slopes and failure events can assist the design work. The design sectors can also be defined based on geometrical considerations. For instance, a slope with a nose or haul ramp requires different stability parameters than a slope with the same geotechnical domain but different geometry. The subdivision of domains into design sectors can reflect control at all levels, from bench scale, where minor structures provide the main control up to the overall slope scale, where a particular major structure may influence a range of slope orientations within a domain. Three dimensional views of the major geological structures that intersect the current pit shell are shown in Fig. 12a–f.

The slope's geometrical design was planned as a series of batters separated by berms. This original design has been substantially reviewed since the mining operation began. A decrease in the overall slope angles was necessary during the first stage of excavation to eliminate slope instabilities.

Because leaching occurs below the surface and the excavation intersects major faults and many lesser faults and shears, it would be impracticable to assume that steep slopes can be used in the Handlebar Hill open pit design (Rosengren 2007). Realistic interamp and bench slope angles with pit slope design parameters resulting from the design work are specified in Table 5.

The 24 m high bench faces are applicable only on the west slope, which contains the fresh rock. The 12 m slope height is appropriate for oxidized rock mass, and 9 m wide berms are specified throughout because it will be very difficult to maintain berm crests in the weak and highly structured rock. The berms are not narrower than 9 m because the potential for failure is likely to be higher for a narrow rather than a wide berm or slope toe.

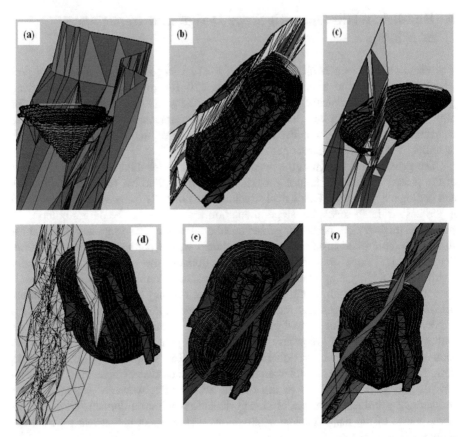

Fig. 12. Major geological structures expressed in three dimensional against the Handlebar Hill open pit shell, (a) Barkly shear fault, (b) Paroo fault, (c) Offset fault, (d) Hangingwall, (e) SBS shear and (f) 4–6 shear

Table 5. General parameters of slopes (after Rosengren 2007)

East and West Slopes	Oxidised	Leached	Fresh
Bench height (m)	12	16	24
Bench slope (°)	65	65	65
Berm width (m)	9	9	9
Interamp slope (°)	39.4	44.2	49.9
North and South Slopes	Oxidised	Leached	Fresh
Bench height (m)	12	16	24
Bench slope (°)	65	70	70
Berm width (m)	9	9	9
Interamp slope (°)	39.4	47.2	53.5

The steeper slopes at the north and the south of the pit reflects the N-S orientation of the bedding and the majority of the faults and shear strikes. These slopes are oriented at right angles to the strike and will have a higher stability. During the two stages of excavation, the lower part of the haul road has been located on the west slope. This design has reduced the interamp and overall slope angles in this zone. At the end of the first stage of excavation, the second haul road loop was located on the toe of the slope. However, the haul road may be less stable because of the leached rock mass located below the haul road, and the angle of the upper slope needs to be reduced.

The details superimposed onto the pit design in Fig. 9 are the interpreted limits of oxidation and leaching rock masses within the pit slopes, and the interpreted geology and structure of the Handlebar Hill open pit is shown in Figs. 4 and 5. The cross-sections of the west wall geometries related to the existing slopes at the west of the pit are modelled by considering the structures and the domains with its final depths (Fig. 13).

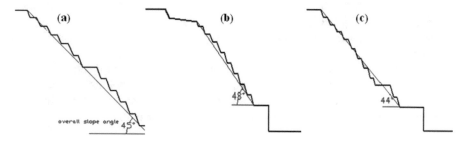

Fig. 13. Cross-sections of the Handlebar Hill open pit west slopes: (a) southern-west slope, (b) middle-west slope and (c) northern west slope

Figure 9 shows that approximately 80% of the slopes for both stages of excavation are positioned in oxidised and leached zones. The fresh rock is exposed in the central part of the west slope and in some fresh upper Urquhart Shale on the middle of the northern slope. This rock mass mainly consists of Shear Siltstone. The whole of the east slope, half of the northern and southern slopes and the pit floors are located within the oxidised or leached zones, which has lower values of rock mass strength parameters.

9   Conclusions

The final slope design of the Handlebar Hill open pit has been reviewed based on data generated from a geotechnical computer program, in-situ slope performance and field investigation. The geotechnical characterisation, underground water resources assessment and stability monitoring addressed issues of the in situ condition. Several key issues that affect the geotechnical design of the final pit have been identified and are as follows:

- The upper part of the west slope is located within oxidised, leached and fresh Eastern Creek Volcanics, with the Paroo Fault dipping steeply back into the slope.
- The Magazine Shale in the pit is exposed in a narrow band at the southern-west corner of the west slope. This has caused unstable areas within the leached Magazine Shale and the benches could be at risk after heavy rainfall.
- The Spear Siltstone is the main element in the centre of the west slope and the upper Urquhart Shale is in the lower part.
- The lower Urquhart Shale covers the entire east slope within stage 1 of the excavation. The central Urquhart Shale forms the northern part of the east slope in stage 2 of the excavation.
- The Barkly Shear zone does not intersect the east slope of both pit stages but is closely adjacent to it. This is an important result because the major fault that forms the hangingwall did not coalesce with the Barkly Shear zone.
- Field investigation of the discontinuities found that the surfaces of the joints and bedding planes are generally smooth, but the surfaces of the faults and shear zones varies from smooth to very rough. Joint spacing is less than 250 mm, (i.e. moderately to closely jointed) and the infills are oxides, calcite or clay. In summary, the rock masses would be of poor to fair surface conditions and of very blocky structure.
- There are slopes, such as the SW slope, that consists of four different geological units that present with different degrees of weathering. The north slope consists of two geological units. This can lead to slope instability and therefore, this zone requires drainage and reinforcement to improve slope stability.

Acknowledgements. The authors would like to thank Glencore Zinc for providing permission to carry out this research and publish this paper. In addition, the authors would like to acknowledge Dr. Ahmed Soliman Principal Geotechnical Advisor, for his support.

References

Australian Groundwater and Environmental Consultants Pty Ltd. (AGE): Handlebar Hill Open Cut Groundwater Assessment. Internal Technical Report on Handlebar Hill Open Cut-George Fisher Mine, reported to Xstrata Zinc Mine, Project No. G1370, Australia, April 2007 (2007)

Bieniawski, Z.T.: Engineering Rock Mass Classifications. Wiley, New York (1989)

Caia, M., Kaisera, P.K., Tasakab, Y., Minamic, M.: Determination of residual strength parameters of jointed rock masses using the GSI system. Int. J. Rock Mech. Min. Sci. 44(2), 247–265 (2007)

Glencore Coal: Xstrata Zinc: Lady Loretta and Handlebar Hill Mines expanding. Glencore Coal Press Release on 1 May 2012, HandlebarHillMinesExpanding.aspx (2012)

Hoek, E., Diederichs, M.S.: Empirical estimation of rock mass modulus. Int. J. Rock Mech. V 43, 203–215 (2006)

Hoek, E., Torres, C.C., Corkum, B.: Hoek-Brown Failure Criterion, 2002nd edn. Rocscience Inc., Toronto (2002)

McKnight, S.: Personal communication (2015)

Medhurst, T.P., Brown E.T.: Large scale laboratory testing of coal. In Jaksa, M.B., Kaggwa, W. S., Cameron, D.A. (eds.) Proceedings of 7th ANZ Conference, Geomech, pp. 203–208. IE Australia, Canberra (1996)

Moshab: Geotechnical considerations in open pit mines guideline. Mines Occupational Safety and Health Advisory Board of the Government of Western Australia, Department of Minerals and Energy, Australia, p. 12 (1999)

Neudert, M.: A depositional model for the Upper Mount Isa Group and implications for ore formation. Ph.D. thesis, School of Earth Sciences, Australian National University, Canberra, Australia (1983)

Read, J.: Personal communication, Melbourne (2011)

Review of Hilton Mine Rock Property Data: A memorandum from Mount Isa Mines Limited, dated in September 25 1981, Reference: RPS/7.s/RES MIN 15.1.3 in Xstrata report, Australia (1981)

Rosengren, K.: Handlebar Hill Open Cut Pit. Internal technical report, Xstrata report in 2011, Mt. Isa, Queensland, Australia (2008)

Rosengren, K.: Proposed Handlebar Hill open cut geotechnical review. Technical report, No. 27010, Mt. Isa, Queensland-4825, Australia (2007)

Seville, R.: Review of Hilton Mine rock property data. Mount Isa Mines Limited, Australia (1981)

Simmons, J.V., Simpson, P.J.: Composite failure mechanisms in coal measures rock masses— myths and reality. In: International Symposium on Stability of Rock Slopes in Open Pit Mining and Civil Engineering, South Africa, July 2006 (2006). The Journal of the South African Institute of Mining and Metallurgy, V. 106, pp. 459–470

Sjöberg, J.: Estimating rock mass strength using the Hoek-Brown failure criterion and rock mass classification—a review and application to the Aznalcollar open pit. Division of Rock Mechanics, Department of Civil and Mining Engineering, Lulea University of Technology, Sweden (1997)

Tarrant, G.C., Lee, M.F.: Mount Isa rock properties. Technical mining report no. RES MIN 60, investigation of rock properties conducted by the authors and Stampton V R., Mt Isa Mines Limited, Mt Isa operations in June 1981, Australia (1981)

Valenta, R.K.: Deformation, fluid flow and mineralisation in the Hilton Area. Mt Isa, Australia. Ph.D. thesis, Monash University, Australia (1988)

Initiation Mechanism of Extension Strain of Rock Mine Slopes

Maged Al Mandalawi[1(✉)], Manar Sabry[1], and Mohannad Sabry[2]

[1] Faculty of Science and Technology, Federation University,
Mt. Helen, VIC, Australia
sp.group@ymail.com
[2] School of Computing, Engineering and Mathematics,
Western Sydney University, Sydney, Australia

Abstract. Slopes in open pits exhibit fracturing around excavations, often initiated by extension strain which results from a combination of principal stresses adjacent to the slope boundaries. This extension strain is commonly described using minimum principal strain or minimum principal stress equations. These equations show that the extension strain can expand and fracturing occurs if the extension strain exceeds a critical value. Anisotropic rock masses with multiple and complex structures increase the potential of the development and coalesce of cracks with pre-existing discontinuities for further potential failure.

This paper presents finite element analysis to model the extensions strain implementing the criterion of Stacey (1981). The distributions of extension strain are predicted around slope of Handlebar Hill open pit mine at Mt Isa, Queensland, Australia. Around the pit wall, fracturing near the excavation boundary is often the result of extension strain of the rock. Through the mining activities, fractures in the slope face can manifest into slabbing and spalling. Extension strain may develop circumferential fractures close to the slope surface, the closer to the excavation perimeter, the more open the cracks. The result of the extension strain distribution simulated in this paper is in accord with failure events observed on site. The numerical modelling and the discussion of this study focused on the prediction of potential fracturing zones within the critical values of the extension strain around the slope.

Keywords: Extension strain · Stacy criterion · Rock slabbing · Numerical modelling

1 Introduction

Extension strain in rock can cause minor cracks (initially microscopic crack in rock) that initiated under tensile or minimum compressive stress. It may extend within the elastic zone and become an observable fracture as a result of changing in-situ stress conditions or natural softening material deformation. Deformation and strain are closely related distinct terms. Deformation describes the complete manifestation from initial to final geometry of rock, the strain describes the alterations of points in rock relative to each other.

© Springer Nature Switzerland AG 2020
A. Bezvijen et al. (Eds.): GeoMEast 2019, SUCI, pp. 52–64, 2020.
https://doi.org/10.1007/978-3-030-34178-7_5

Laboratory tests detected two main types of fractures in brittle materials including the extension fractures or so-called opening modes and shear fractures (Ramsey and Chester 2004). Other approaches with more experimental studies on rocks showed that a pre-existing fracture does not extend as a single flaw within the rock mass. For example, Brady and Brown (2006) reported that fractured rock should include a large number of very minor extensional fractures that developed before the process of the final propagating.

The transition of extension strain from tensile to shear fracture as being a function of confining stress by a unique laboratory uniaxial compression tests was described by Hoek (1968). It illustrates the estimated fracture condition for rock including the extension of the strain from the initiation, until the complete rock fracturing.

Stacey and de Jongh (1977) reported that the Griffith and Mohr-Coulomb criteria in terms of fracturing initiation were ineffective to predict the extent of fractures associated with underground boring in hard rock mass. An equation to calculate the extension strain magnitudes of different rocks was required. The equation has to be effective in predicting the extent of fracturing zones together with the localised orientation of these fractures around slope and underground excavations.

The rock failure criteria included only the major and minor principal stresses in their equations, in which the failure dominated by shear stress. But, in many open pit slopes and underground tunnels it was noticed there were parallel fractures near excavations like spalling or splatting. Ndlovu and Stacey (2007) reported that this rock failure shows tensile nature, but not related to shear fracturing. Several criteria were suggested to determine the rock critical strain e.g. Sakurai (1981) in underground excavation and Stacey (1981) in excavated slopes which accounts for the three principal stresses. Therefore, this criterion included that tensile failure can occur even if the applied stresses are compressive.

In studies focusing on the mechanisms of compound rock failures, Eberhardt et al. (2004) argued that processes relating to extension strain, internal degradation of rock mass strength, brittle fracture damage and shearing are also instrumental in the kinematics of step-path failure in massive rock slopes. Many studies Liang (2003), and Yang et al. (2008) presented results of triaxial extension experiments on rocks show that a continuous transition from extension strain to shear fracture with an increase in confining pressure, which formed under compression. This combined fracture mechanism has been considered under compressive and tensile stress states at critical angles to the maximum principal compressive stress.

Kwasniewski and Takashashi (2010) presented a new strain criterion to calculate the mean normal strain for conditions of low confinement. They found that the highest extension strain (the least principal strain) at strength of failure is changeable under triaxial compression with different values of confining pressure used in triaxial compression laboratory tests.

The source of stress in the slope is a combination of the pre-existing tectonic stresses and the stress reduction normal to the slope as well as stress concentration in some locations of the slope. The extensional fracturing occurs when the intermediate and major principal stress components are large enough or the minor principal stress is small enough to result in minor principal strain component to be extensional and in which exceeds a critical value.

Fracturing possibly occurs at the stage that the bench is a part of the floor, ramp or the toe of the slope. Conditions for extensional fracturing are expected within the benches in the slope when they are close to the haul road and the toe. Rock masses of fresh quality, which is existed at the west slope of the open pit, would be more susceptible to extensional fracturing.

Extension strain can be described by all three principal stresses, therefore, suggesting that tensile strain will occur when all stresses are compressive. This criterion proposed that the extension can occur through indirect tension if the tensile strain goes beyond a value, which is reliant on the condition of the rock units. While, the failure criteria developed by many authors, e.g., Mohr-Coulomb (Coulomb 1779) and generalised Hoek-Brown criteria (Hoek et al. 2002) included only major and minor principal stresses in their equations involving the rock failures by shear deformation.

Due to its development and flexibility, the finite element stress method is progressively being applied to slope stability analysis (Mahtab and Goodman 1970) and (Stacey et al. 2003). Non-linear finite element models using elastic-perfectly plastic material strength have been used to obtain factor of safety of slopes, which offers a number of advantages over general limit equilibrium method (Hammah et al. 2004).

This paper studies the application of finite element stress method to the determination of the extension strain of rock slope in open pit for which rock strength is modelled by the generalised Hoek-Brown failure criterion. The paper first utilises the standard finite element method used for slope stability analysis of Hoek-Brown slopes. It then demonstrates the distributions of extension strain through application of the method, and uses the user defined data dialog to inputs mathematically the extension strain equation of Stacy.

One of the major advantages of Phase^2 finite element slope stability is the analysis parameters can be customised if required. Using the user define data after the interpretation of the finite element model to illustrate contour of user data. This code is able to create a new user data definition for variable inputs with plotting results of defined equations to solve complex non-linear problems.

Figure 1 shows a view of the Handlebar Hill open pit at Mt Isa, Queensland, Australia. The extension strain and the influence of their distribution on the stability of the pit slope stability are modelled using the finite element stress method implementing the criterion of Stacey (1981) of principal stresses.

Fig. 1. View of Handlebar Hill open pit, photo taken from the south in February 2011

2 Review of Extension Strain Criterain Proposed by Stacey (1981)

The extension strain criterion is a fracturing initiation criterion and not a shear strength criterion of ultimate strength. It explains that the confining stress does not cause the fracturing. Confining stress always works against the generation of rock fractures. The extension strain criterion, in fact, states that it is the lack of confining stress will result in extension strain within elastic limit of rock material. Once the extension strain exceeds a critical value, tensile fracturing can develop plastically. Therefore, the high level of confining stress detects the threshold of rocks, grain size porosity and mineralogy. This will inhibit the initiation of extension strain and, subsequently, the developments of cracking events.

Stacey (1981) described the first empirical criterion for extension strain and fracturing initiation of brittle rock. It was a new method proposed to obtain the critical value of extensions strain from many laboratory tests by plotting axial strain against lateral strain. He applied his criterion in two cases: in an open cut pit for haulage road side-wall bulging and in underground mine for rock spalling. His work proved that the extension strain could occur when the major principal stresses are compressive stresses. The initiation of a fracture in the rock mass will occur in a direction normal to the extension strain and corresponds to the least compressive stress path, when the strain exceeds a critical value, which is rock type reliant. The expression of this criterion is as follows: $e > e_c$ where (e_c) is the critical or limiting value of extension strain for the rock. The extension strain may expand in different directions within rock slopes. When strain exceeds a critical value, which is a characteristic of that rock mass, imitation of extension strain will occur and the developments of these fractures are likely (Stacey 1981).

However, the critical magnitudes of extension strain in rock masses could be evaluated by this criterion. This way of obtaining the extension strain can be designated crack initiation by strain as this corresponds to the stage at which the crack starts within the rock. Louchnikov (2011) stated that it has to be noted that the first term of the extension strain criterion was limiting tensile strain criterion, hence emphasising the tensile nature of the extension strain.

The criterion of Stacy for a linear elastic material is an out of plane strain analysis to define the minimum principle strain (σ_3) which calculates the three principal stresses by using the following three dimensional elastic equations:

$$\varepsilon_3 = [\sigma_3 - \upsilon(\sigma_1 + \sigma_2)]/E \tag{1}$$

Where σ_1, σ_2 and σ_3 are the three principal stresses

ε_3 = Extension strain, which is the strain in the direction of the minor principal stress
υ = Poisson's ratio
E = Modulus of elasticity.

This criterion allows defining the three-dimensional principal stress directions, including stress, displacement and strength factor and, therefore, the results are more accurate compared to the plane strain assumption. Stead et al. (2007) stated that Stacey et al. (2003) and Stacey (2006) used 3D elastic extension strain criterion in the analysis of large scale open cut slopes to present the significance of considering the extension strain as a potential rock fracture. Ndlovu and Stacey (2007) reported that this extension strain criterion can be implemented for a wide range of rocks, from strong, brittle to soft rock such as coal. Furthermore, extension strain is not limited to deep open pits and high slopes, the effect will be more noticeable in brittle rock condition in which fractures can develop rapidly (Stacey, personal communication, June 2 2015).

The extension strain criterion can be described in three dimensional strain space within a cubic state boundary surfaces condition as illustrated in Fig. 2. The inner cube presents the extension strain initiation and the outer cube describes the manifestation to crack damage strain. The principal strains are defined in terms of in plane by ε_1 and ε_3, and the out of plane strain by ε_2 as shown in the cubic strain spaces. Principal strains are ε_1, ε_3 and ε_2 (major, minor and intermediate principal strain) are oriented with respect to stresses redistribution by excavation. In the three dimensions cubic, ε_3 is not necessarily the minor principal strain, if the value of ε_2 is less than ε_3 at the same location, then the in plane ε_3 is the three dimensional intermediate principal strain.

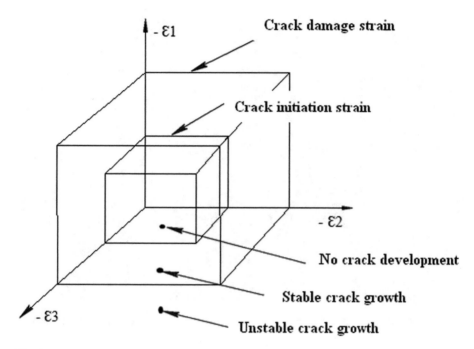

Fig. 2. Extension strain failure criterion in three dimensional cubic strain spaces (after Wesseloo 2000)

3   In Situ Observations

The following examples of rock failures occurred at Handlebar Hill open pit, where the extension strain expected to be developed and initiated the slope face failures. These failures included rock spalling from the face of the slope (Fig. 3), and sidewall slabbing below the ramp (Fig. 4), which involved a thin extent of detached rocks.

The fractures of rocks will progress as far as allowable by the distribution of extension strains, and their existence inhibited by the compressive confining stress. It follows that the presence of extension strain is likely to be limited to the proximity of openings in stressed rock. This happens in excavated locations of rocks and, hence, prediction of rock fracture is necessary (Stacey 1981). The extension strains then developed to rock fractures, and these fractures propagated to slope faces failure, and slabbing in the sidewall on the least compressive stress extent.

Martin et al. (1999) stated that slabbing and splatting in a rock may occur in a stable manner or violently in the form of tensile strain bursts in underground tunnels having low confining pressure similar to that at an excavated slope. These slabs can range in thickness from a few millimetres to several square metres in surface area, and the formations and thicknesses of these slabs could be related to strain energy.

The propagation of cracks extends into rock masses and may coalesce through pre-existing discontinuities during stress loading. In a heterogeneous slope, tensile cracks that propagate in their own plane and out of plane are normal to confining pressure

Fig. 3. Sidewall spalling in the face of the pit bench, photo taken from the south in February 2011

Fig. 4. Side-ramp failure (slabbing) at the open pit ramp, photo taken from the south in Feb. 2011

direction (the minimum principle stress). Tensile cracking could be initiated, also, by compressive stress. Therefore, adding compressive stress on the rock masses, the possible shear movement on rock plane with less strength zones would be a potential mode of tensile failure. Hoek (1968) indicated that the tight connection of asperities can increase the differential shear stress displacement that results in the propagation of existing vertical cracks, which lead to vertical tensile rupture.

3.1 The Source of Extension Strain in Open Pit and the Orientation of Potential Fractures

The increases of confining stress inhibit fracture propagation and should induce a move of the rock towards the ductile field (Valenta 1988). In an underground mine when rock is exposed to an applied differential stress in deeper excavation, the tight connection of asperities and increased differential shear stresses can arise. The geological structure where dilation is normal to the sliding surface is restricted by adjacent blocks under high confining stresses. This confining stress is presented by several hundred meters of overlying rock resisting block mobilisations.

In open cut mines, all of the earth's crust is in a state of three dimensional stresses. When an excavation is made, it changes the stress distribution locally and usually there is some stress concentration around the openings. It is these stresses that contribute to the confinement, or lack of it around the excavation. With regard to open pits, the geometry of the slopes in plan and section will influence the redistribution of stresses. The materials of the geological units can also influence local stresses, for example, stronger beds will attract more stress than the weaker or more ductile layers, which could lead to localised failure. In such cases, stress redistribution will extend deeper into the rock mass unaffected by blasting. However, at low confining stress, rock will be brittle and tends to crack sooner (Nelson 2003).

Extension fractures in rocks form in planes normal to the direction of the extension strain that corresponds with the direction of the minimum principal stress. In excavated slopes the directions of the minimum principal stresses are normal to the confining pressure of slope and because of the direction of the confining pressure being normal to the slope surface. Therefore, the orientations of fractures are likely to be sub-parallel to the slope surface behind the slope face and the inclined extension fractures can propagate near the slope toe. The simulated levels of extension strain contours make the development of extension fractures possible and that could considerably reduce slope stability.

3.2 Extension Strains Around Two Dimensional Numerical Models

This section describes the numerical modelling of the extension strain zones and tensile failure around the first and the second stage of slope excavation. The existence of extension strain indicates that the rock masses have the potential to expand in one or more directions in the slope. Critical extension strain will be calculated according to Stacey criterion (1981), which proposes that rock fracture can be developed from extension strains that exceed the critical value of −0.0003 or greater. The pit depth is 180 m, overall slope angle of 45° and (K) ratio, which is the horizontal to the vertical in situ stresses ratio values of 0.6 for in-plane and 0.86 for out of plan stress ratio, a value of 0.2 for Possion's ratio, and rock mass modules of 8,248.32 MPa.

Two dimensional finite element stress elastic analysis modelled the results using Phase² software. The simulation of the two excavation stages slope shows zones of extension strain near the slope toe and the floor of the pit. Zones of extension strain occur alongside the lower benches of the slope and upper benches located above the ramp. The extension strain is concentrated around the benches located above the ramp and behind the slope crest (Fig. 5).

The model in Fig. 6 illustrates the development of strain near the toe and along the floor surface, and zone of strain extend behind the crest of the first stage of excavation.

Although the extension strain values obtained are low, due to the limited height of pit slope, the extension strain will act normal to the foliation planes and stimulating extension fracturing (opening up deformations) on these rock planes (Stacey et al. 2003). These predicted extension strains in the pit, at any stage of excavation may change through mining operation to develop more fractures.

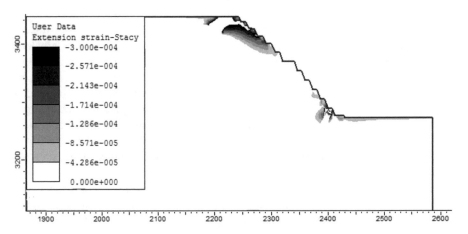

Fig. 5. Extension strain distribution around slope of 45° angle and 180 m height of the two stages of excavations

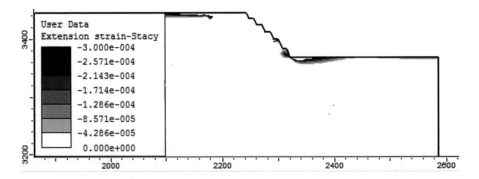

Fig. 6. Extension strain distribution around slope of 45° angle and 180 m height of one stage of excavation

3.3   Tensile Stress around Two Dimensional Numerical Models

The condition of stress distribution is conductive to the propagation of the extension fracturing adjacent to the slope. The two dimensional analyses allowed modelling the tensile stress distributions and magnitudes around the slope. Zones of tensile stress are concentrated near the crest and adjacent to the upper benches of the slope of two stages of excavation as shown in Fig. 7. The tensile stress distribution within the first stage of slope excavation of 80 m height is shown in Fig. 8. The tensile stress is concentrated near the upper benches, but no tensile zones presented near the toe of both stages of slope excavation.

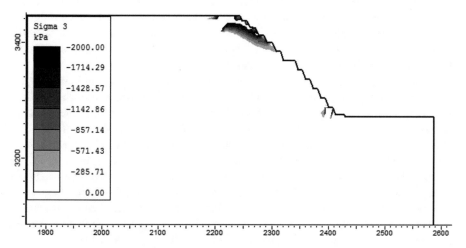

Fig. 7. Tensile stress distributions around slope of 45° angle and 180 m height of two stages of excavations

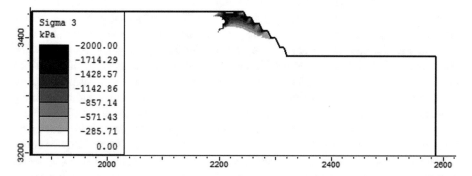

Fig. 8. Tensile stress distribution around slope of 45° angle and 80 m height of one stage excavation

3.4   Combined Failure Mechanism in Open Pit Slopes

Several types of slope failures are often attributed to the development of the extension strain have other causes. In open cut pit slopes, extension strain can be considered as a contributor to slope failures. Complex rock failure involves mechanisms related to extension strain initiation around a slope, which propagates to the fracturing of intact rock and the deformation along geological structures. An example of large scale rock failure involving the extension strain in a complex mechanism of failure is the two massive rock-slides of the 1991 Randa town cliff in Switzerland. While some geotechnical investigations did not refer to the role of extension strain in the complex failure mechanisms in Randa rock failure, Sartori et al. (2003) did not identify whether these series of rock mass failure events combined, acted as an exceptional failure trigger. On the other hand, Eberhardt et al. (2004) pointed out that the results showed that the rock slope failure of Randa involved the initiation and propagation of brittle tensile fractures driven by extensional strain that interacted with natural pre-existing discontinuities to form internal shear planes.

If the role of extension strain is over emphasised, then this may lead to missing the real cause of slope failure. This suggests that more than one mechanism of rock behaviour may contribute to final slope failure. This failure is a function of combined mechanisms, namely the extension strain imposed around excavations, may, manifest itself and notch expansion through faults and joints. It is the strain, which is stored within the elastic limit of the rock. Because of the low confining pressure around slopes of an open pit, the extension strain will progress and extend to one or more directions and develop a network fracturing in the rock masses. This extension may interact with the natural large scale discontinuities to form a critical failure in the slope surface and increase slope instability. The extension strain may coalesce with minor or major geological structures in rock mass and this will possibly lead to a localised slope failure (Louchnikov 2011).

Joint sets often have deep impacts on elastic and strength properties in rock masses and then on open pit slope stability (Pariseau et al. 2008). In slopes of jointed rock mass, the extension strain is influenced remarkably by the joint orientations. Then, the joint spacing can coalesce with the opening of extension strains, and propagates within the orientations of joints. In heterogeneous rock excavations subjected to stresses, tensile cracks may not propagate in their own plane, but they may propagate near the joint surface and spacing. The extension strain that progresses to brittle fracturing in intact rock will be associated with the opening of joints. Hence, this provides the deformation of rock along the orientations of joints, bedding, geological contacts or any other discontinuities. Significant cohesion loss could occur along this path, which possibly could result in the formation of different scales of blocks, which disadvantages the slope stability.

As confining pressure decreases in excavated slope, rock permeability increases. This explains the common incidence of reducing porosity and permeability of rocks with the pit depth in sedimentary basins (England et al. 1987). Joint spacing and cracks opening escalating in response to the decreasing of applied normal stress (Brown and Scholz 1986). Rock permeability increases at highly strained slopes due to micro-cracking of extension strain at low effective pressure, therefore, slope stability could be sensitive to the growth of pre-existing micro-cracks, joints, and faults relative to the principal stresses.

4 Conclusions

This paper discussed the modelling of extension strain in the slope of Handlebar Hill open pit at Mt Isa, Queensland, Australia. The extension strain around two dimensional finite element elastic models is applicable using the criterion of Stacey (1981). This paper demonstrates that extension strain is the most probable starting point for the fracturing mechanisms. It was shown that the extension strain develops around the toe presenting in a larger magnitude behind the crest and is limited in the upper benches.

There is an agreement on the influence of the extension strains on slope stability between the in-situ observations and the predicted fracturing distributions of this study. As presented, the out-of-plane principal stresses extension strain criterion assist in understanding the probable initiation and orientation of extension fractures in the pit rock slope.

Acknowledgements. Authors wish to thank Prof. T. Stacey for kindly providing his hypothetical observations on the rock failure mechanisms occurring at the pit slopes, and being available for helpful comments.

References

Brady, B.H.G., Brown, E.T.: Rock mechanics for Underground Mining, 3rd edn. Springer, Netherlands (2006)

Brown, S.R., Scholz, C.H.: Closure of random elastic surfaces in contact. J. Geophys. Res. **90**, 5531–5545 (1986)

Coulomb, C.A.: Essai sur une application des règles de maximis et minimis a quelques problemes de statique. relatifs al architecture, Memoires de mathematique et de physique, presentes al Academie Royale des Sciences par divers savans, vol. 7, pp. 343–382 (1779)

Eberhardt, E., Stead, D., Coggan, J.: Numerical analysis of initiation and progressive failure in natural rock slopes – the 1991 Randa rockslide. Int. J. Rock Mech. Min. Sci. **41**(1), 69–87 (2004)

England, W.A., MacKenzie, A.S., Mann, D.M., Quigley, T.M.: The movement and entrapment of petroleum fluids in the subsurface. J. Geol. Soc. Lond. **144**, 327–347 (1987)

Hammah, R.E., Curran, J.H., Yacoub, T.E., Corkum, B.: Stability analysis of rock slopes using the finite element method. In: Proceedings of the ISRM Regional Symposium EUROCK 2004 and the 53rd Geomechanics Colloquy, Salzburg, Austria (2004)

Hoek, E.: Brittle failure of rock-in rock mechanics in engineering practice. In: Stagg, K.G., Zienkiewicz, O.C. (eds.), pp. 99–124. Wiley, London (1968)

Hoek, E., Carranza-Torres, C., Corkum, B.: Hoek-Brown failure criterion. Edition. In: Proceedings 5th North American Rock Mechanics Symposium, pp. 267–273. University of Toronto Press, Toronto (2002)

Kwaśniewski, M., Takahashi, M.: Strain-based failure criteria for rocks: state of the art and recent advances. In: Zhao, J. et al. (ed.) Rock Mechanics in Civil and Environmental Engineering, pp. 45–56. CRC Press/Balkema, Leiden, Netherlands (2010)

Liang, Z.Z.: Three-dimensional Numerical Modelling of Rock Failure Process, Doctoral Thesis, Dalian University of Technology (2003)

Louchnikov, V.: Simple calibration of the extension strain criterion for its use in numerical modelling. In: Potvin, Y. (ed.) Strategic versus Tactical Approaches in Mining 2011. Australian Centre for Geomechanics, Perth, Australia (2011)

Mahtab, M.A., Goodman, R.E.: Three-dimensional finite element analysis of jointed rock slopes. In: Proceedings of the Second Congress of the International Society of Rock Mechanics, Belgrade, vol. 3, pp. 353–360 (1970)

Martin, C.D., Kaiser, P.K., McCreath, D.R.: Hoek–Brown parameters for predicting the depth of brittle failure around tunnels, Application of the in plane minimum principal strain criterion. Geotech. **36**, 136–151 (1999). Canada

Ndlovu, X., Stacey, T.R.: Observations of roof guttering in a coal mine. In: Proceedings of the 3rd Southern African Rock Engineering Symposium, Best Practices in Rock Engineering, Randburg, S. Afr. Inst. Min. Metall., Symposium Series S41, 2005. pp. 285–300 (2007)

Nelson, A.S.: Deformation of Rock, Lecturer in Physical Geology EENS 111, Tulane University, Louisiana, U.S.A (2003)

Pariseau, G.W., Purib, S., Schmelterc, C.S.: A new model for effects of impersistent joint sets on rock slope stability. Int. J. Rock Mech. Min. Sci. **45**(2), 122–131 (2008)

Ramsey, M.J., Chester, M.F.: Hybrid fracture and the transition from extension fracture to shear fracture, Center for Tectonophysics, department of Geology and Geophysics, Texas A & M, University, College Station, Texas, USA (2004)

Sakurai, S.: Direct strain evaluation technique in construction of underground opening. In: Preceding of the 22nd U.S. Symposium on Rock Mechanics, 29 June–2 July, Cambridge, MA, pp. 278–282 (1981)

Sartori, M., Baillifard, F., Jaboyedoff, M., Rouiller, J.D.: Kinematics of the 1991 Randa rockslides. Nat. Hazards Earth Syst. Sci. V **3**(5), 423–433 (2003)

Stacey, T.R.: A simple extension strain criterion for fracture of brittle rock. Int. J. Rock Mech. Min. Sci. **18**, 469–474 (1981)

Stacey, T.R., De Jongh, C.L.: Stress fracturing around a deep level bored tunnel. J. S. Afr. Inst. Min. Metall. **78**, 124–133 (1977)

Stacey, T.R.: Personal communication (2014–2015)

Stacey, T.R., Terbrugge, P.J., Keyter, G.J., Xianbin, Y.: A new concept in open pit slope stability and its use in the explanation of two slope failure. In: Fifth Larger Open Pit Mining Conference, 3–5 November, Kalgoorlie, WA, pp. 259–266 (2003)

Stead, D., Coggan, J., Elmo, D., Van, M.: Modelling brittle fracture in rock slopes: experience gained and lessons learned. In: Australian Centre for Geomechanics' International Symposium on Rock Slope Stability in Open Pit and Civil Engineering, Perth, Australia, pp. 239–252 (2007)

Valenta, K.R.: Deformation, fluid flow and mineralization in the Hilton area, Mt. Isa, Australia, Doctoral Thesis of Philosophy in the department of earth science, Monash University, Victoria, Australia (1988)

Wesseloo, J.: Predicting the extent of fracturing around underground excavations in brittle rock, Technical report, SRK Consulting, Johannesburg, South Africa (2000)

Yanga, Q.S., Jianga, Z.Y., Xu, W.Y., Chen, X.Q.: Experimental investigation on strength and failure behavior of pre-cracked marble under conventional triaxial compression. Int. J. Solids Struct. **45**(17), 4796–4819 (2008)

Stress in Rock Slopes: Non-persistent Versus Persistent Joint Model

Abdel Kareem Alzo'ubi[1(✉)] and Osama Mohamed[2]

[1] Abu Dhabi University, Al Ain, UAE
abdel.alzoubi@adu.ac.ae
[2] Abu Dhabi University, Abu Dhabi, UAE

Abstract. In rock slopes prone to topple, mechanical, physical or chemical weathering might degrade the rock mass properties and compromise the stability of that rock slopes. However, the nature of the joints distribution and the degree of continuity is extremely important in determining the stresses magnitude and sign inside the rock masses. This is mainly due to the stress concentration and distribution near the joint tips inside the rock slope if rock bridges exist. In this paper, a man-made rock slope is numerically modeled and investigated under two types of joint sceneries; out-dipping persistent joint and non-persistent joint models. One point, near the crest, inside the rock slope were used to monitor and measure the major and minor principal stresses as the excavation at the toe of the slope progressed step by step. To generate the two models, the discrete element method with the Voronoi tessellation joint pattern was utilized. One of the advantages of this modeling approach is that it al-lows for generating non-persistent joint as well as persistent joint patterns inside the rock slope. The results of this numerical study show that in large rock slopes such as the one examined in this study, the stress concentration in the non-persistent model is three to four times larger than the stresses at the same point in the persistent joints model. The minor principal stresses in both modeling cases were negative tensile stresses and up to several MPa in magnitude.

1 Introduction

In natural or man-made rock slopes, many discontinuities such as joints, flaws, and faults exist to weaken the rock mass strength. These rock discontinuities can be either persistent or non-persistent joints. This degree of continuity affects the stress magnitudes and type inside the rock slopes (Alzo'ubi 2016a). Alzo'ubi (2016a) discussed thoroughly factors affecting rock slope movement styles, some of these factors are; rock mass characteristics, joints orientation, and joints degree of continuity. This paper discusses a rock slope, susceptible to toppling movement, to study the effect of degree continuity on the stress magnitude and type inside large rock slopes where the confinement stresses are not high as in the underground environment. To investigate this issue, two geological settings, of the same rock slopes, were examined; a first model with persistent joint set and the second model with a non-persistent joint model.

In the 1970's De Freitas and Watters (1973), presented toppling to describe rock slope instability where rock columns bend forward. Later, Goodman and Bray (1976),

© Springer Nature Switzerland AG 2020
A. Bezvijen et al. (Eds.): GeoMEast 2019, SUCI, pp. 65–72, 2020.
https://doi.org/10.1007/978-3-030-34178-7_6

showed three movement styles associated with toppling; flexural, block toppling, and block-flexural toppling. More recently, Alzo'ubi et al. (2010) showed that toppling is mainly controlled by tensile strength as long as sliding is allowed between the rock columns that dip into the slope. How-ever, they did not discuss the type of stresses in rock masses containing non-persistent rock joints. This paper introduces the difference between large rock slopes with persistent joints and large rock slopes with non-persistent joints.

The High Valley Copper (HVC) mine is located in British Columbia, it consisted of two pits; the Lornex Pit and the Valley Pit. The rock slope in this paper is an open pit mine is the southeast wall of the Lornex pit. This slope was discussed by Tosney (2001) by using conventional discrete element approach to create the model and to simulate the two sets of joints as continuous ones. Moreover, he assumed a tensile strength of 0.1 MPa for all the rock units' constitute the rock slope. Probably, he used this low value to have an acceptable agreement between the displacement in the numerical model and the displacement in the field. In addition, he did not consider the non-persistent nature of the rock joints out-dipping of the open pit slope.

The southeast wall of the Lornex pit at the HVC was also modeled by using the UDEC-DM (Alzo'ubi et al. 2007). This UDEC-DM is utilized in this paper because it enables the modeler to create persistent and non-persistent joint sets and allows for rock mass fracturing once shear and/or tension exceeds the rock mass strength. More importantly, this added degree of freedom in the discrete element method allows for stress concentrations to develop freely around joint tips of non-persistent rock joint model. This stress concentration in the rock bridges area may cause tensile stresses to rise and cause intensive fracturing as we will see in the following sections.

2   Geological Setting

In this case history, the geological investigation of this rock slope revealed two major sets of joints; a persistent set of joints that dips into the slopes at angles between 70° and 80° from the horizontal with a spacing between 20 and 40 m. And a second non-persistent joint set that dips in the direction of the slope face at angles between 50° and 60° with spacing between 10 m and 20 m. The bridge length between the non-persistent joint ranges between 15 and 20 m. This paper aims to model the geological model; the persistent joint set and the non-persistent one by using a discrete element damage model developed by (Alzo'ubi 2009). This model allows for stress concentration to evolve and initiate shear/tensile fractures inside a rock mass.

This slope suffers from toppling in the range of meters. In order to determine the rock mass properties as shown in Fig. 1, the engineers utilized the RMR classification system (Bieniawski 1976). Furthermore, Tosney 2001 used the Hoek-Brown criterion (Hoek and Brown 1988) to determine the Mohr-Coulomb strength properties to be used to conduct the numerical analysis (Table 1). The site engineers assumed that the tensile strength, of all rock units forming this pit, is 0.1 MPa, which is at the low end of the tensile strength range for such rock units, in reality, this value can be up to 5 MPa. The low tensile strength value of 0.1 MPa was used probably to facilitate the stability analysis of the slope by using conventional discrete element modeling (Tosney 2001).

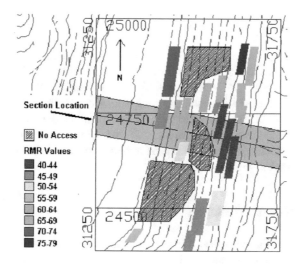

Fig. 1. The RMR classification of the pit around the selected section (Tosney et al. 2004)

Table 1. Mohr-Coulomb strength properties based on the RMR rating used

RMR 76	φ (°)	C (MPa)	E (GPa)	G (GPa)	K (GPa)
40	29	0.7	5.6	2.2	4.7
60	43	1.0	17.8	6.8	14.8
80	54	2.1	56.2	21.6	46.9

The numerical simulation in this study, modeled the slope in two different modes to investigate the stress concentration; the first model assumes persistent out-dipping joint set and the second model assumes non-persistent out-dipping joint set. The major principal stress and the minor principal stress were monitored and compared. Table 2 shows the properties of the two joint sets that are used in these simulations.

Table 2. The in-dipping and out-dipping joints properties used in the simulation, Alzo'ubi et al. (2012)

Parameter	In-dipping faults	Out-dipping joints
Orientation (°)	70	125
Normal Stiffness (GPa/m)	4	4
Shear Stiffness (GPa/m)	1	1
Cohesion (kPa)	6.0	0
Friction (°)	12	25

3   Displacement and Movement Modes

To maintain safe operation at the mine, the engineers installed a monitoring system capable of capturing the displacement location as well as its rate. More details about this system can be found at Newcomen et al. (2003), I utilized these displacements to verify the numerical model by comparing the numerical results along with the actual field measurements as the mining operation continued at the pit. The background movement, not related to the mining steps, was subtracted from the total displacement. Three modes of movements were observed at this section (Alzo'ubi 2009); toppling, sliding, and bulging at the toe of the slope.

4   The Numerical Models

Alzo'ubi et al. (2007), developed a discrete element approach called the UDEC-DM, this modeling procedure simulates the rupturing of rock masses through generating flaws inside the rock mass. The approach can simulate two models; persistent joint pattern as well as non-persistent joint pattern. The Lornex pit southeast wall is modeled in this paper by using the two joint patters to quantify the major-minor principal stresses.

The details of the model were discussed by Alzo'ubi et al. (2012); the general the geometry of the model and the actual mining steps conducted in the mine is shown in Fig. 2. To generate the mine two joint sets as discussed above, two models were created as shown in Figs. 3 and 4. The first model (Fig. 3) consists of two joint sets; the in-dipping joints with an angle of 70° and spaced at 30 m and the out-dipping persistent joint set at an angle of 55° from the horizontal and spaced at 15 m. In the second model (Fig. 4); the first joint set is the in-dipping joints with an angle of 70° and spaced at 30 m, while the second out-dipping joint set is a non-persistent one at an angle of 55° from the horizontal and spaced at 15 m. The bridge length between the joint tips was set at 20 m.

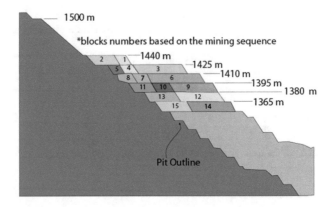

Fig. 2. Blocks excavation sequence at Lornex southeast wall (from Alzo'ubi 2009)

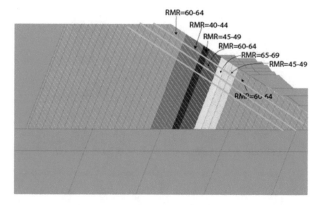

Fig. 3. The persistent joint model along with the RMR classification

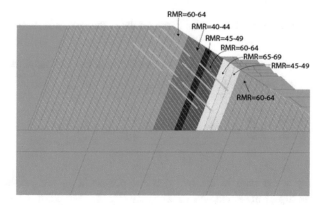

Fig. 4. The non-persistent joint model along with the RMR classification

The area susceptible to toppling was divided by the Voronoi pattern with a 1.4 m edge length. This Voronoi joint pattern created polygonal blocks, see the inset of Fig. 5.

During the numerical simulation, blocks 1 to 15 were mined while monitoring the slope displacement and compared the results with the field measurements. The mining-induced deformation was monitored by Slope Monitoring Prisms (SMP) at the location of the chosen section. To calibrate the model, it was brought to a stable condition by using high strength properties. Following that, the model was assigned realistic strength properties and was calibrated. The displacement of SMP #413 (Alzo'ubi 2009) was utilized at two mining stages. To get better agreement between the field and numerical displacements, the normal and shear stiffness were varied until a good agreement was achieved at the properties shown in Table 3.

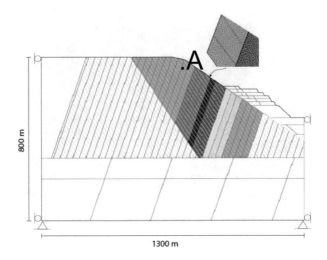

Fig. 5. The UDEC-DM and the details of the generated flaws

Table 3. Normal and shear stiffness resulting from the calibration process

RMR 76	Normal Stiffness (GPa/m)	Shear Stiffness (GPa/m)
40–44	11	3.5
45–49	14	4.5
60–64	34	11

To study the stress path in the two models, persistent vs non-persistent models, point A near the top of the slope was chosen, see Fig. 5. This point is located near the joint in the persistent joint model and near the joint tip in the non-persistent model. Each block (Fig. 2) were excavated numerically, the minor and major principal stresses were recorded at equilibrium (were displacement became constant).

5 Results and Discussion

The simulation started by the excavation of blocks 1 to 15, one block at a time while monitoring the rock mass fracturing and more importantly the stress changes at point A. The results revealed that stress concentration was much higher in the non-persistent model once compared with the persistent model. This behavior was observed by Alzo'ubi (2016b) in the finite element method as well. As shown in Fig. 6, the major principal stresses at the two models showed similar values, however, the minor principal stress in the non-persistent joint model is 4 times the magnitude of the minor principal stress in the persistent model. The minor principal stress in the two models was tension as shown in the Figure. This shows that tensile stresses will develop in the model and will cause tensile fracturing as observed by Alzo'ubi (2016a, b). Tensile strength has a major role in the stability of rock slopes but also more important if the

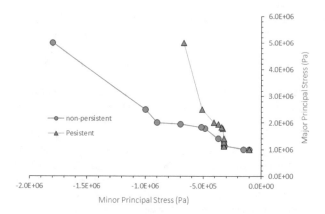

Fig. 6. Major and Minor Principal Stresses in the two models at point A

joint sets are non-persistent ones. In the case were tensile stresses exceeds the tensile strength of the material fracturing will occur in either model. In the field, the tensile strength of the material needs to be accurately evaluated and the geological model need not be simplified by assuming a persistent joint model while the actual one is non-persistent.

6   Conclusions

In this research, a rock slope was modeled by using two geological models; the first one with persistent joint model and the second one with the non-persistent joint model. The UDEC-DM was utilized to generate the two models, this modeling approach is capable of modeling persistent and non-persistent joint sets. The major and minor principal stresses were monitored as the mining process occurred at the toe of the mine. Although the minor principal stresses in both models were negative, the stress concentration magnitude at the non-persistent model was almost four times higher than the negative stress magnitude in the persistent model. This shows that tensile strength in of the intact rock can be as high as several mega Pascal and the stress concentration at the joint tip in rock masses can fracture the rock.

Acknowledgments. The author would like to acknowledge the financial contribution of Abu Dhabi University and the ORSP under grant number 19300344.

References

Alzo'ubi, A.K.: State of the art, reconstruction of damaged zones: transitory stresses effect and factors controlling rock mass stability. Int. J. Geomate **14**(41), 35–43 (2018)

Alzo'ubi, A.K.: Rock slopes processes and recommended methods for analysis. Int. J. Geomate **11**(25), 2520–2527 (2016a)

Alzo'ubi, A.K.: Modeling yield propagation of jointed synthetic rocks. In: Ulusay, et al. (eds.) Rock Mechanics and Rock Engineering: From the Past to the Future, Proceeding of the 2016 ISRM International Symposium, EUROCK 2016, Cappadocia, Turkey, 29–31 August 2016, vol. 1, pp. 113–118. CRC Press: Taylor & Francis Group, London (2016b). ISBN 978-1-138-03265-1

Alzo'ubi, A.K., Martin, C.D., Cruden, D.: A discrete element damage model for rock slopes. In: Eberhardt, E., Stead, D., Morrison, T. (eds.) Rock Mechanics, Meeting Society's Challenges and demands, Vancouver, B.C., Canada, vol. 1, pp. 503–510 (2007)

Alzo'ubi, A. K., Martin, C.D., Cruden, D.M.: Effect of mining on deformation patterns in a large open pit rock slope. In: 21st Canadian Rock Mechanics Symposium, RockEng 12-Rock Engineering for Natural Resources, Edmonton, Canada, pp. 229–236 (2012)

Alzo'ubi, A.K., Martin, C.D., Cruden, D.M.: Influence of tensile strength on toppling failure in centrifuge tests. Int. J. Rock Mech. Min. Sci. **47**, 974–982 (2010)

Bieniawski, Z.: Rock mass classification in rock engineering. In: Balkema, A. (ed.) Exploration for Rock Engineering, Cape Town, Johannesburg, vol.1, pp. 97–106 (1976)

De Freitas, M., Watters, R.: Some field examples of toppling failure. Geotechnique **23**, 485–514 (1973)

Goodman, R., Bray, J.: Toppling of rock slopes. In: Specialty Conference on Rock Engineering for Foundations and Slopes, pp. 201–234. ASCE, Boulder (1976)

Hoek, E., Brown, E.: The Hoek–Brown failure criterion – a 1988 update. In: Proceedings of the 15th Canadian Rock Mechanics Symposium, Department of Civil Engineering, University of Toronto, Toronto, Canada. pp. 31–38 (1988)

Newcomen, W., Murray, C., Shwydiuk, L.: Monitoring pit wall deformation in real time at Highland Valley Copper. In: CAMI Conference (2003)

Tosney, J., Chance, A., Milne, D., Amon, F.A.: Modelling approach for large-scale slope instability at Highland Valley Copper. In: Mining Millennium 2000: International Convention and Trade Exhibition. Content Management Corp. Richmond Hill, ON, Toronto, Canada (2000)

Tosney, J.: A design approach for large scale rock slopes. Master's thesis, University of Saskatchewan, Saskatoon (2001)

Tosney, J., Milne, D., Chance, A.V., Amon, F.: Verification of a large scale slope instability mechanism at Highland Valley Copper. Int. J. Surf. Min. Reclam. Environ. **18**(4), 273–288 (2004)

Waldner, M., Smith, G., Willis, R.: Lornex. In: Brown, A. (ed.) Porphyry Deposits of the Canadian Cordillera, vol. 15. CIM (1976)

Load Capacity of Helical Piles with Different Geometrical Aspects in Sandy and Clayey Soils: A Numerical Study

Amir Akbari Garakani[1(✉)] and Jafar Maleki[2]

[1] Niroo Research Institute (NRI), Tehran, Iran
aakbari@nri.ac.ir
[2] Geotechnical Engineering, Sharif University of Technology, Tehran, Iran

Abstract. In recent years, using helical piles as deep foundations for different types of structures has been increased considerably. In this paper, by using finite element software (ABAQUS), the compressive and tensile load capacities of helical piles screwed in sandy and clayey soils have been studied numerically and corresponding load-displacement curves are presented. For this purpose, different geometrical aspects of the helical pile element (including the pile length, the main shaft diameter, the helix diameter and the internal helix spacing) have been taken into account for different soil properties conditions. In modeling efforts, a disturbed zone around the pile element is also considered for better catching the effect of the soil disturbance during pile installation procedure.

Based on the obtained results, it is observed that for both types of the studied soils, increasing the helix diameter leads to an increase in load capacities. In addition, it is observed that by increasing the internal helix spacing up to three times of the helix diameter, the ultimate load capacities were increased and then remained almost unchanged. Similar trend was obtained from parametric study on the main shaft diameter. However, increasing the pile length was shown to have consistent increasing effect on the ultimate load capacities.

In this paper, to verify the obtained results, some experimental records are also considered and compared statistically with corresponding load capacities from numerical simulations. Comparisons show very good agreement between the numerical results, the experimental records and analytical solutions.

Keywords: Helical pile · Numerical study · Geometrical parameters · Sand · Clay · Experimental verification

1 Introduction

Implementing deep foundations are necessary when the subsurface layers are weak to carry the superstructure loads, and in the depths, there is a soil with higher resistance. Wooden and metal piles, cast-in-place concrete piles, driven piles, and helical piles are several common types of deep foundations. In recent years, using helical piles, as a kind of deep foundation, have been increased significantly. These piles are usually installed by imposing a driven torsion moment at the pile head. One of the most

© Springer Nature Switzerland AG 2020
A. Bezvijen et al. (Eds.): GeoMEast 2019, SUCI, pp. 73–84, 2020.
https://doi.org/10.1007/978-3-030-34178-7_7

significant advantages of helical piles is the capability of their installation under desired dip angle. In addition, they can provide high resistance against tensile forces compared with another types of deep foundations.

Initial views on evaluation of the helical pile capacity included the testing of the behavior of shallow helixes. Some researchers with physical model testing showed an ideal disconnection cone in an experiment of a helical pile (Mors 1959; Turner 1962). Physical modeling by Ghaly et al. (1992) indicated overloading on the sand surface results in increasing the load capacity of the helical piles that by increasing the burial depth, it will be diminished. Rao and Prasad (1993) studied the final load capacity of helical anchors in terms of the shear strength parameters of the soil and indicated that the helix spacing affects the load capacity, firmly.

Livneh and El Naggar (2008) compared the full scale test results of 19 numbers of the helical piles load tests by conducting finite element modeling with PLAXIS 3D and suggested that the final compressive capacity of the helical piles should account for a pile settlement equal to 8% of the helix diameter plus the elastic deformation of the pile shaft. Salhi et al. (2013) investigated the effect of helix diameter and spacing on the final load capacity by performing 2D FEM modeling with ABAQUS and reported that the optimized ratio of the helix spacing to the helix diameter (S/D_p) varies between 1.5 to 2. In addition, it was observed that helixes with smaller diameters usually had more considerable effects on the pile capacity. George et al. (2017) examined the load capacity of the helical piles with different S/D_p in driven in loose clayey soils with different densities by performing numerical simulations using PLAXIS 3D software. Results showed that increasing the relative density of the soil leads to an increase in compressive and tensile ultimate load capacities. Furthermore, the load capacity increases by increasing the installation depth and shaft diameter, which the former has much considerable influence.

In this paper, a parametric study has been performed on the compressive and tensile ultimate bearing capacities of the helical piles by conducting both FE numerical simulations using ABAQUS and implementing an analytical solution. Accordingly, load-displacement curves have been obtained from FE analyses as well as the ultimate bearing capacities form analytical solution. Then, by implementing the results from field tests in sandy and clayey soils, the numerical and analytical results are verified. Results showed good agreement between the predicted and the experimental load bearing capacity values.

2 Method of Analysis

The soil parameters and the geometrical aspects have the most important roles on the load capacity of the helical piles (Garakani 2019). In current research, to examine the aforementioned factors, several numerical simulations have been conducted on helical piles with several geometric features using a FE software, ABAQUS.

During installation of the helical piles, due to shaft penetration and helix collision with the surrounded soil, a disturbed zone is created around the pile element (Fig. 1). In modeling efforts in this research, this disturbed zone is modeled by assuming a cylinder-shaped region having a diameter of two times of the helix diameter and a length equal to the pile length. In disturbed zone, soil parameters (i.e., frictional angle, cohesion and elastic modulus) were reduced in accordance with Buduh (2011).

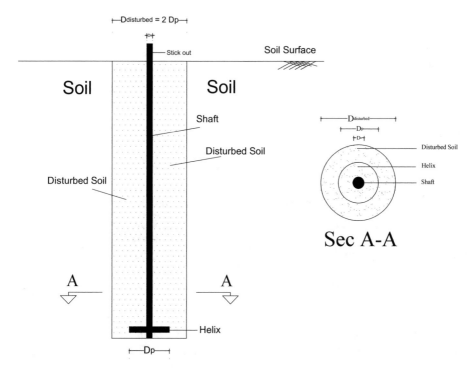

Fig. 1. A schematic view of the helical pile, disturbed and undisturbed soil zone

In addition, typical sandy and clayey soil types were assessed with different geometrical aspects of the pile. For a helical pile driven in undrained saturated clayey soil, the water table was considered at the ground surface, and the rate of the imposed load was considered as fast to invoke the undrained soil condition. Accordingly, the Poisson's ratio of 0.49 was assumed for the analysis of the saturated clay (Elsherbiny and El Naggar 2013). Further, the adhesion between the pile and the clayey soil was taken as 25 kPa (Canadian Foundation Engineering Manual 2006). In this paper, the sandy and clayey soils were modeled by considering Mohr-Columb (MC) constitutive framework. Accordingly, the elastic behavior is described by Poisson's ratio, v, and Young's modulus, E, and the plastic behavior of the soil is defined by taking into account the residual friction angle, ϕ_r, the dilation angle, ψ. Also, material hardening is defined by the cohesion yield stress, c. In Table 1, soil parameters are shown.

To investigate the role of the geometrical aspects of the helical piles on the compressive and tensile bearing capacity values, the pile length (H), the shaft diameter (D_s), the helix diameter (D_p), and the internal helix spacing (S) were examined. These parameters are given in Table 2.

Table 1. Soil parameters used in this study

Soil parameters	Undisturbed sandy soil	Disturbed sandy soil	Undisturbed saturated clayey soil	Disturbed saturated clayey soil
Residual friction angle (ϕ_r)	33	27	–	–
Dilation angle (ψ)	5	0	–	–
Cohesion (kPa)	–	–	60	50
Adhesion (kPa)	–	–	30	25
Young's modulus (Mpa)	30	15	35	35
Unit weight (kN/m^3)	18	17	18	18

Table 2. Summary of helical piles geometric parameters

Pile length (m)[ft]	Shaft diameter (in)[mm]	Helix diameter (in)[mm]	Internal helix spacing (in)[mm]	Helix numbers
4, 8, 12 [13.1, 26.3, 39.4]	6, 9, 12, 15 [152.4, 228.6, 304.8, 381]	12, 18, 24 [304.8, 457.2, 609.6]	24, 48, 72 [609.6, 1219.2, 1828.8]	3

In this study, a displacement equal to 5% of the helix diameter is considered as the corresponding pile displacement at failure as suggested by Elsherbiny and El Naggar (2013).

3 Verification of the Numerical Simulations

The numerical modeling procedure and obtained results should be verified by considering the field-test data, the laboratory testing data, or the analytical solution results. Therefore, firstly, the field test results under compressive loading for two helical piles in sandy and clayey soils was chosen from Elsherbiny and El Naggar (2013). Studied piles were executed in two sites: site (A) that is located in northern Alberta, Canada, and is mainly composed of sand; and site (B) that is located in northern Ontario, Canada, and represents clay soil primarily. Two axial compressive load tests were conducted at sites (A) and (B). The soils properties for sites (A) and (B) and the piles properties are presented in are shown in Tables 3 and 4, respectively. Additional descriptions on these field testing are detailed by Elsherbiny and El Naggar (2013). The load-displacement curves for numerical modeling and field testing data in the sites (A) & (B) are shown in Fig. 2. It is observed that numerical results are in good agreement with the load test results respecting the soil properties, the helical pile characteristics for piles PA-1 and PB-1.

Table 3. Soil parameters of site A and B

Site name	A		B	
Depth soil (m)	0–5	5–9	0–3	3–7
Description	Sand (Compact)	Sand (Compact)	Sandy silt (Compact)	Silty Clay (Very Soft)
ϕ_r (deg)	24	21	–	–
ψ (deg)	10	10	–	–
C_u (Kpa)	–	–	36	9
C_a (Kpa)	–	–	34	9
Friction factor	0.44	–	–	–
γ (kN/m^3)	20	20	17	17
k_S	0.55	0.55	1.0	1.0
ν	0.3	0.3	0.49	0.49
E (Mpa)	50	50	24	7

Table 4. Tested piles configurations for site A & B

Site name	Pile depth (m) [ft]	Stick out (m) [ft]	Number of helixes	Helix diameter (in) [mm]	Shaft diameter (in) [mm]	Installation torque (KN-m)
A	5.5 [18]	0.3 [1]	1	24 [609.6]	10.75 [273]	25.9
B	7.2 [23.6]	0.4 [1.3]	3	24 [609.6]	7 [177.8]	9.5

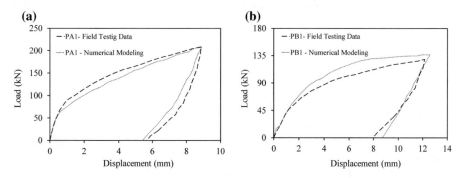

Fig. 2. Load-displacement curve for a calibration modeling, (a) PA-1 pile in sandy soil, and (b) PB-1 pile in clayey soil

Secondly, a comparison of ultimate loading capacity between field-testing data, FEM results, and analytical solutions were done (Fig. 3). By two approaches can determine the ultimate bearing capacity of a helical pile; i.e., individual and cylindrical shear method. Based on the individual capacity method, ultimate bearing capacity, P_{u1}, is the sum of individual bearing capacities of n helical bearing plates plus adhesion along the shaft, provided by

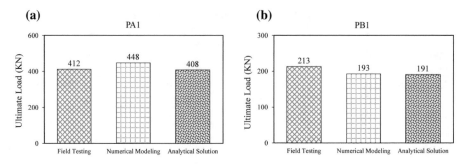

Fig. 3. Ultimate load capacity of helical piles by field testing, FEM and analytical solution methods

$$P_{u1} = \sum_n q_{ult}A_n + \alpha H (\pi d) \tag{1}$$

Where

q_{ult} is the ultimate bearing pressure
A_n is the area of the nth helical bearing plate
α is adhesion between the soil and the shaft
H is the length of the helical pile shaft above the top helix, and
d is diameter of a circle circumscribed around the shaft.

Also, ultimate bearing capacity of a helical pile according to the cylindrical shear method, P_{u2}, is determined by considering the sum of shear stress along the cylinder, adhesion along the shaft, and bearing capacity of the bottom helix provided by

$$P_{u2} = q_{ult} A_1 + T (n - 1) s \pi D_{avg} + \alpha H (\pi d) \tag{2}$$

where:

A_1 is the area of the bottom helix, T is the soil shear strength, H is the length of shaft above the top helix, d is the diameter of the pile shaft and, the term $(n - 1) s$ is the length of soil between the helices.

Finally, bearing capacity of a helical pile is a minimum of P_{u1} and P_{u2}. The perfect spacing of helices is corresponded to the condition that the bearing capacity results from individual and cylindrical shear methods become the same. To calculate the bearing capacity based on above equations, some researchers have presented relations for the implemented parameters. In this paper, Perko (2009) equations were used as primary considerations for both individual and cylinder failure mechanisms. Figure 3 shows that the ultimate compressive capacity has a good fitting for all three cases (namely, analytical, numerical and experimental cases).

4   Results

Numerical simulations were performed under 3D condition, and long-term behavior of the soil was considered. The mesh dimensions are chosen in such a way that the domain and mesh size geometry do not influence the obtained results. Figure 4 shows a schematic mesh pattern for the constructed model.

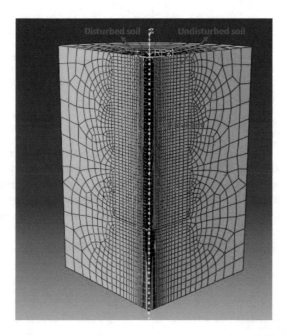

Fig. 4. A schematic configuration of the F.E mesh for modeling a helical pile problem

4.1   Compressive Loading Simulations

Pile Length Effect
The pile length effects on helical piles compressive capacity are examined in three different lengths of 4, 8, and 12 m, in sandy and clayey soils (Fig. 5). Generally, in helical piles, by increasing the pile length, the ultimate loading capacity increases. That is because of the increasing in the buried part length of the pile and consequently, the increase in the load capacity of the pile shaft due to the increase in the frictional contact surface with the soil. It is also observed that for the undrained saturated clay, the piles have reached to the ultimate state under about 10 mm total displacement, while this value for sandy soils slightly increases.

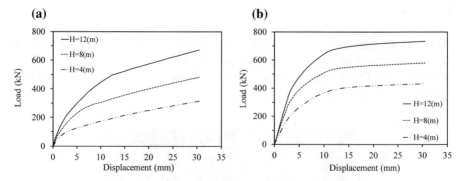

Fig. 5. Load-displacement curve for a helical pile on compressive loading in three pile lengths, (a) sandy soil, and (b) clayey soil

Shaft Diameter Effect

Data presented in Fig. 6 reveal that the shaft diameter has a considerable influence on the pile compressive capacity. In sandy soil, it is observed that as the shaft diameter increases, the load capacity is increased significantly. However, by changing the shaft diameter form 12 to 15 in, there is almost no significant change observed. By examining the contours of the shear stress around the pile, it can be noticed that this parameter remains almost constant through the buried depth of the pile (Fig. 7). Figure 7 also indicated that by increasing the shaft diameter, the resilient force has increased.

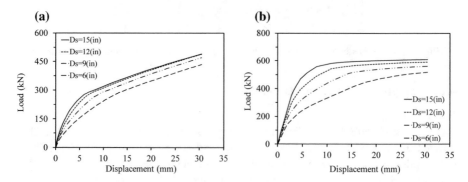

Fig. 6. Load-displacement curve for a helical pile on compressive loading in four shaft diameter, (a) sandy soil, and (b) clayey soil

(a)   **(b)**

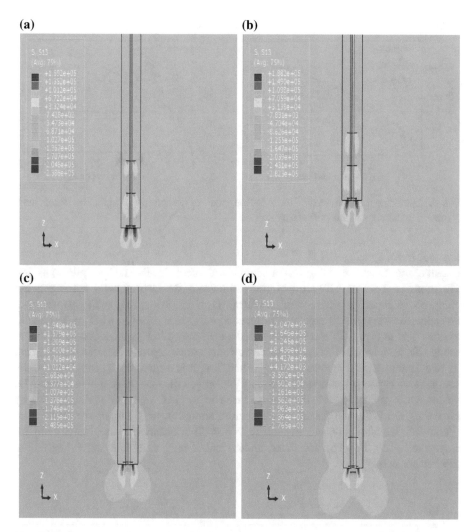

(c)   **(d)**

Fig. 7. Shear stress contour for a helical pile on compressive loading in sandy soil for shat diameter of, (a) 6 in, (b) 9 in, (c) 12 in, and (d) 15 in

Helix Diameter Effect

To assess the effects of the helix diameters on the compressive capacity of a helical pile, a numerical case with H = 8 (m), D_s = 15 (in), and S = 72 (in) has been examined with three different helix diameters. The results are shown in Fig. 8. It is observed that by increasing the helix diameter (D_p), the compressive load capacity increases as well in both sandy and clayey soils. In both soil types, in the range of small displacements, the load capacity values are almost the same, however, by increasing the pile tip displacements, variations are much considerable. The reason is that at smaller displacements, the pile capacity is controlled by the friction along the soil cylinder formed around the helixes and highly dependent on D_s, but at large displacements, failure mechanism is ruled by the bearing capacity of the helixes.

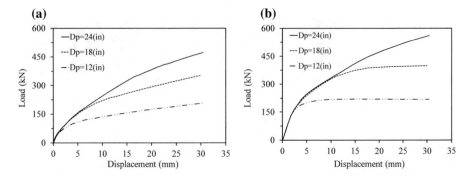

Fig. 8. Load-displacement curve for a helical pile on compressive loading in three helix diameter, (a) sandy soil, and (b) clayey soil

Internal Helix Spacing Effect
Usually, the factor of S/D_p has a significant effect on the helical pile capacity. Therefore, in this study, for studying the influence of the internal helix spacing on the compressive capacity, a case with $H = 8(m)$, $D_s = 15$ (in) and $D_p = 24$ (in) has been investigated with three different condition of $S = 24$, 48 and 72 (in) ($S/D_p = 1$, 2 and 3) as illustrated in Fig. 9. In accordance with the data shown in Fig. 9a, it is noticed that in sandy soil, by increasing S/D_p, the ultimate compressive capacity increases as well, but its rate decreases with increasing S/D_p. Therefore, it is estimated that in $S/D_p = 4$, the compressive capacity has remained almost unchanged. Its main reason is in lower S/D_p values where the failure mechanism occurred in form of a "cylinder shape," but by increasing the S/D_p, the failure mechanism is almost of an "individual failure" case (Perko 2009; Elsherbiny and El Naggar 2013). This observation is almost valid for clayey soils, too (Fig. 9b). However, under smaller S/D_p values the individual failure mechanism is happened.

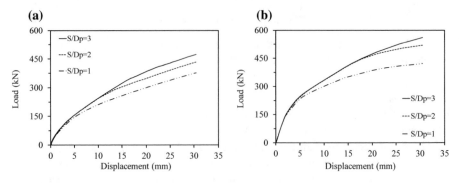

Fig. 9. Load-displacement curve for a helical pile on compressive loading in three S/D_p, (a) sandy soil, and (b) clayey soil

4.2 Tensile Loading

In some previous researches, it is mentioned that the tensile capacity of the helical pile is almost 85% of its compressive capacity (Elsherbiny and El Naggar 2013; Perko 2009; Livneh and El Naggar 2008; Tappenden and Sego 2007). In addition, it is frequently reported that the helical pile length has a significant effect on the tensile load capacity. Accordingly, in this study, the helical pile length effects on the tensile loading capacity has been analyzed, as the results are depicted in Fig. 10. Figure 10 shows that in both sandy and clayey soils, by increasing the pile length, the pile tensile capacity increases, similar to compressive load capacity. Also as shown in Table 5, it is remarked that the ratio of the tensile loading capacity to the compressive loading capacity for the helical piles studied in this research varies between 0.78 to 0,96 that shows a good agreement with previous studies. However, in clayey soil this ratio is obtained close to unity for both load conditions.

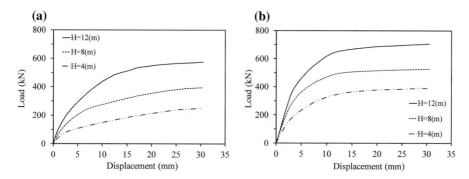

Fig. 10. Load-displacement curve for a helical pile on tensile loading in three pile lengths, (a) sandy soil, and (b) clayey soil

Table 5. Ratio of the tensile loading to the compressive loading capacity

Pile length (m)	Sandy soil	Undrained saturated clayey soil
4	0.78	0.90
8	0.82	0.92
12	0.86	0.96

5 Conclusions

In this paper, by implementing a Finite Element Software, ABAQUS, a parametric study has been carried out on some geometrical aspects of the helical pile elements driven in sandy and undrained saturated clayey soils. Accordingly, corresponding features (consisting pile length, shaft diameter, helix diameter, and internal helix spacing) have been taken into account. The obtained results from numerical modeling showed that:

1. By increasing the helical pile length, in both studied soils, the ultimate compressive capacity increases.
2. Increasing the shaft diameter leads to an increase in ultimate load capacity, however, its effect diminishes by reaching the shaft diameter.
3. It is noted that by increasing the S/D_p up to 3, the ultimate load capacities were raised and then almost remain unchanged.
4. By increasing the helix diameter, similar trends have been obtained from the parametric study and the ultimate load capacity has been increased.
5. Tensile loading capacity of a helical pile is obtained as (80% to 90%) and (90% to 100%) of the compressive loading capacity in sandy and undrained saturated clayey soil conditions, respectively.

Some experimental records are also scrutinized and compared with corresponding load capacities from numerical and analytical solutions. Comparisons show perfect agreement between the field tests, FEM results, and analytical solutions.

Acknowledgments. This research has been financially supported by Niroo Research Institute (NRI) of Iran, which is greatly acknowledged.

References

Buduh, M.: Soil Mechanics and Foundations, 3rd edn. Wiley, Hoboken (2011)

Canadian Foundation Engineering Manual, 4th edn. Canadian Geotechnical Society (2006)

Elsherbiny, Z.H., El Naggar, M.H.: Axial compressive capacity of helical piles from field tests and numerical study. Can. Geotech. J. (2013). https://doi.org/10.1139/cgj-2012-0487

Garakani, A.: A guideline on implementing helical piles in 63 kV towers. Technical report, Niroo Research Institute, Tehran, Iran (2019)

George, B.E., Banerjee, S., Gandhi, S.R.: Numerical analysis of helical piles in cohesionless soil. Int. J. Geotech. Eng. (2017). https://doi.org/10.1080/19386362.2017.1419912

Ghaly, A., Hanna, A., Ranjan, G., Hanna, M.: Helical anchors in dry and submerged sand subjected to surcharge. J. Geotech. Eng. **117**, 1463–1470 (1992)

Livneh, B., El Naggar, M.H.: Axial testing and numerical modeling of square shaft helical piles under compressive and tensile loading. Can. Geotech. J. (2008). https://doi.org/10.1139/T08-044

Mors, H.: The behaviour of mast foundations subjected to tensile forces. Bautechnik **36**(10), 367–378 (1959)

Perko, H.A.: A Practical Guide to Design and Installation. Wiley, New Jersey (2009)

Rao, S.N.R., Prasad, Y.V.S.: Estimation of uplift capacity of helical anchors in clays. J. Geotech. Eng. **119**, 352–357 (1993)

Salhi, L., Nait-Rabah, O., Deyrat, C., Roos, C.: Numerical modeling of single helical pile behavior under compressive loading in sand. Electron. J. Geotech. Eng. **18**, 4319–4338 (2013)

Tappenden, K., Sego, D.: Predicting the axial capacity of screw piles installed in Canadian soils. In: The Canadian Geotechnical Society (CGS), Ottawa Geo 2007 Conference, pp. 1608–1615 (2007)

Turner, E.: Uplift resistance of transmission tower footing. J. Power Div. **88**(2), 17–34 (1962)

Strength and Deformation Characteristics of Laterite Rock with Different Rock Matrix's Using Triaxial System

M. V. Shah[✉] and Prashant Sudani

LD College of Engineering, Ahmedabad, India
drmvs2212@gmail.com, sudaniprashant93@gmail.com

Abstract. Strength and deformation behavior of jointed rock is essential in the viewpoint of design of slopes, construction of tunneling and mining projects due to the presence of natural hair cracks, fissures, bedding planes, faults, etc. In order to avoid problems like loss of strength due to cracks during construction, strength and deformation characteristics of various rock matrices is helpful to simulate such cracks and to give a reliable solution. For this purpose, experimental based study was conducted on laterite rock procured from Kutch, Gujarat to evaluate the shear parameters of the proposed rock mass matrix having two horizontal cuts (at H/4 and H/2 height of the specimen) with 0°, 10° and 30° inclination with horizontal. Stress-strain characteristics and strength behavior was studied using Mohr-Coulomb strength theory and results were compared with intact rock and various jointed rock matrix specimens. Results derived from Mohr-Coulomb strength theory were also compared with the Hoek-Brown strength theory. Results demonstrated that the orientation of the rock matrix shows little increase in cohesion value and decrease in angle of internal friction with increasing inclination of cut from zero to 30° with respect to horizontal. Failure pattern was observed in jointed rock matrix specimens comprises of axial failure, block rotation, splitting of blocks through joints and shear failure.

1 Introduction

Rock differs from most other engineering materials as it contains discontinuities such as joints, bedding planes, folds, sheared zones and faults which render its structure discontinuous. For practical purposes, rock mechanics is mostly concerned with a rock on the scale that appears in engineering, mining and tunneling work, and so it might be regarded as the study of the properties and behavior of accessible rock due to changes in stresses or other conditions. Singh and Rao (2016) studied the strength and deformation behavior of the jointed rock matrix. Arzua et al. (2014), Shah and Patel (2017) studied shear parameter of intact and different rock matrix form specimen. Shah and Joshi (2016) analyzed the effect of the full circular opening with varying sizes on a shear parameter of sandstone. The behavior of discontinuities between the intact rocks is a subject of investigation of many researchers since it is not possible to deal with the same theoretical considerations as intact rock materials (Huang et al. 2015). It is probable that the dilation angle will be at its maximum value just at the instant when peak strength is passed. From the experimental investigations on non-planar joints,

© Springer Nature Switzerland AG 2020
A. Bezvijen et al. (Eds.): GeoMEast 2019, SUCI, pp. 85–102, 2020.
https://doi.org/10.1007/978-3-030-34178-7_8

it has been observed that at low normal stress, a non-planar joint continues to dilate with increasing shear displacement but at reduced angle, while at high normal stress a non-planar joint might cease to dilate altogether after passing its peak strength, if the normal stress is very high, dilation may not take place at any stage (Barton 1973). However, very less amount of experimental study was carried out in past in this direction especially for understanding the shearing behavior of rocks. Thus this research work has provide better understanding in this field especially related to cracks and joint openings for intact rock mass and one with various jointed rock matrix. Scope of the study includes conducting triaxial axial test of cylindrical rock specimen under laboratory environment for various confining pressure and constant strain rate to study the actual shear parameters and behavior of failure pattern with stress and axial strain characteristics of intact rock and various jointed rock matrix. Strength behavior of jointed rock matrix were then compared with the strength of intact rock specimen. Stress-strain behavior were also compared with Mohr-Coulomb strength theory and Hoek-Brown strength theory. Laterite types of rock was used for present investigation.

2 Sampling

Samples were prepared from the chunk of the rock which was procured from the Kutch district, Gujarat, India. Cylindrical samples were cut out from chunk of the procured rock with help of the core cutting machine in L.D. College of engineering Ahmedabad, Gujarat. The core cutting of cylindrical rock core was carried out as per IS-9179-2001 and cylindrical specimens with aspect ratio 1:2 was used to make intact and different type of rock matrix patterns.

Various types of matrixes used in this experimental investigation were as shown in Fig. 1. Rock cutter machine was used to cut the intact rock with various matrix form viz., I) Intact rock specimen with l/d ratio as a 2, II) two inclined cut at H/4 distances from both end with 30° inclination (2I_H/4_30°) as shown in Fig. 2, III) two inclined cut at H/2 distances from both end with 30° inclination (2I_H/2_30°) as shown in Fig. 3, IV) two inclined cut at H/4 distances from both end with 10° inclination(2I_H/4_10°) as shown in Fig. 4, V) two inclined cut at H/2 distances from both end with 10° inclination (2I_H/2_10°) as shown in Fig. 5, VII) Two inclined cut at H/4 distances from both end with 0° inclination (2I_H/4_0°) as shown in Fig. 6, IX) Two inclined cut at H/2 distances from both ends with 0° inclination (2I_H/2_0°) as shown in Fig. 7.

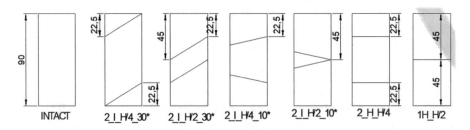

Fig. 1. Various rock matrix pattern

Fig. 2. 2I_H/4_30° rock matrix

Fig. 3. 2I_H/2_30° rock matrix

Fig. 4. 2I_H/4_10° rock matrix

Fig. 5. 2I_H/2_10° rock matrix

Fig. 6. 2I_H/4_0° rock matrix

Fig. 7. 2I_H/2_0° rock matrix

To join cutting pieces of the cylindrical specimen, cement paste was used as binding material to assemble rock block matrix. The end surface of the samples were kept flat and smooth which would helpful for making the good contact between loading pad of testing machine. Assembling of the rock matrix has been done with the use of portland cement paste. Cement paste used for assembling the matrix has 40% of water content. Jointed matrix form was kept in water for curing purpose for 28 days (Fig. 8).

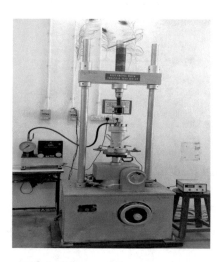

Fig. 8. Electronic rock triaxial setup

3 Methodology

3.1 Index Properties

Index properties for rock specimen includes specific gravity, void ratio, dry density, water absorption and water content were performed as per IS: 13030:1991.

3.2 Rock Triaxial Test

The rock triaxial test was performed by electronic rock triaxial testing machine as per IS 13047:2010. It mainly consists of closed triaxial cell, a loading frame and a confining pressure generating device as shown in Fig. 12 and the shear parameters were obtained from experimental study. Series of tests were performed with different confining pressures varying from 1 MPa, 2 MPa, 3 Mpa & 4 MPa for intact and different rock matrix patterns specimens of laterite rock at the constant strain rate of 0.115 mm/min. Different types of confining pressure were generated using constant pressure system.

4 Theoretical Development

4.1 Mohr Coulomb Strength Theory

The shear parameters of intact and jointed rock specimens were calculated by fallowing equation.

$$\tau = \sigma tan\emptyset + c$$

The major principle stresses of intact and jointed rock specimens were calculated by Mohr-Coulomb strength theory.

$$\sigma_1 = \frac{2c \cdot cos\emptyset}{1 - sin\emptyset} + \frac{\sigma_3(1 + sin\emptyset)}{1 - sin\emptyset}$$

Where,

σ_1 = Major principle stress
σ_3 = Major principle stress
c = cohesion
ϕ = Angle of internal friction

4.2   Hoek-Brown Strength Theory

The Generalized Hoek-Brown failure criterion for jointed rock masses is defined by:

$$\sigma_1 = \sigma_3 + \sigma ci\left(\left(mi \cdot \frac{\sigma_3}{\sigma ci}\right) + s\right)^{0.5}$$

Where,

'σ_1' and 'σ_3' are the major and minor principal stresses at failure, 'mi' is the value of the Hoek-Brown constant 'm' for the rock mass, 's' is constants which depend upon the rock mass characteristics, and 'σ_{ci}' is the uniaxial compressive strength of the intact rock pieces.

In order to use the Hoek-Brown criterion for estimating the strength and deformability of jointed rock masses, three properties of the rock mass have to be estimated. These are: (i) uniaxial compressive strength σ_{ci} of the intact rock pieces, (ii) Value of the Hoek-Brown constant mi for these intact rock pieces, (iii) Value of the constant S for rock.

5   Experimental Investigation

5.1   Properties of Laterite Rock

The cylindrical specimen of intact rock and different rock matrix having 45 mm diameter and 90 mm height were obtained as per IS 13030-1991 and tested to evaluate physical properties of intact laterite rock specimen. Table 1 shows physical properties of intact rock.

Table 1. Properties of laterite rock

Properties	Values
Dry Density (gm/cm^3)	3.18
Porosity (%)	7.66
Unconfined compressive strength (Mpa)	21.5
Water content (%)	0.592
Specific gravity	3.50
Void ratio (%)	8.29
Water absorption	2.726

5.2 Results and Analysis

Triaxial test results in terms of shear parameter (c and ϕ) and modulus of elasticity are demonstrated in Table 2. From Fig. 9, it was observed that the percentage decrement in cohesion (c) for various rock matrixes 2I_H/4_30°, 2I_H/2_30°, 2I_H/4_10°, 2I_H/2_10°, 2I_H/4_0°, and 2I_H/2_0° were 18.68%, 10.49%, 5.24%, 29.50%, 43.60%, and 17.37% respectively with respect to L_I rock. As shown in Figs. 10, 11, the value of angle of internal friction (ϕ) for various matrixes 2I_H/4_30°, 2I_H/2_30°, 2I_H/4_10°, 2I_H/2_10°, 2I_H/4_0°and 2I_H/2_0° was found to be decreased by 24.5%, 26.43%, 18.35%, 20.81%, 10.02%, and 4.60% respectively with respect to L_I rock. The increase in strength of various horizontal cuts specimens were moderate as with increase in confining pressure because the rock blocks came more close to each other at high confining pressure and it requires more normal load to deform them. The percentage increment in modulus of elasticity (E) for 2I_H/4_30°, 2I_H/2_30°, 2I_H/4_10°, 2I_H/2_10°, 2I_H/4_0°, and 2I_H/2_0° were 17.28%, 12.44%, 7.60%, 5.76%, 0%, and 0.57% respectively with respect to L_I rock as presented in Fig. 12. The modulus of elasticity of jointed rock matrix specimens were increases as the numbers of cut increases. The modulus of elasticity decreases as the confining pressure increases for the same number of horizontal joints.

Table 2. Measured values of cohesion and angle of internal friction of different types of rock matrices

Matrix pattern with notation	Shear parameters		Modulus of elasticity
	C (Mpa)	Φ (°)	E (Mpa)
Intact rock specimen, L_I	3.05	48	8.68
2-Horizontal cut with 30° inclination at H/4, 2I_H/4_30°	2.48	36.24	10.18
2-Horizontal cut with 30° inclination at H/2, 2I_H/2_30°	2.73	35.31	9.76
2-Horizontal cut with 10° inclination at H/4, 2I_H/4_10°	2.89	39.19	9.34
2-Horizontal cut with 10° inclination at H/2, 2I_H/2_10°	2.15	38.01	9.18
2-Horizontal cut with 0° inclination at H/4, 2I_H/4_0°	1.72	43.19	8.67
2-Horizontal cut with 0° inclination at H/2, 2I_H/2_0°	2.52	45.79	8.73

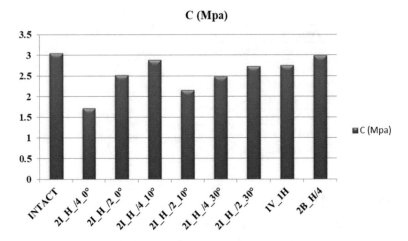

Fig. 9. Comparative plot, cohesion vs. types of matrix as compare to intact.

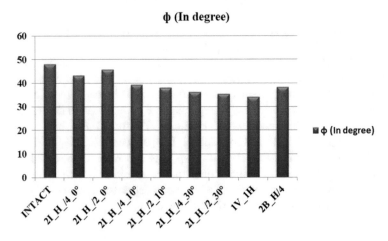

Fig. 10. Comparative plot, angle of internal friction vs. types of matrix as compare to intact.

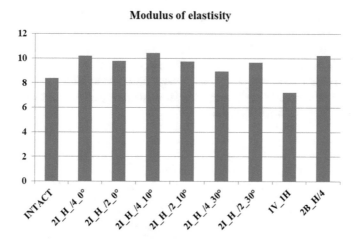

Fig. 11. Comparative plot, modulus of elasticity vs. types of matrix as compare to intact.

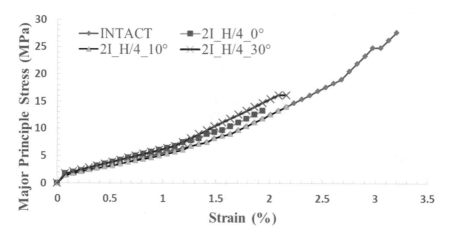

Fig. 12. Comparison of stress-strain of jointed rock matrix pattern (two inclined cut at H/4 distance from both end with inclination of 0°, 10°, and 30°) at 1 Mpa

6   Stress Strain Characteristics

Stress-strain characteristics of the rock matrix were compared with the intact rock specimen. Results depicts that with constant confining pressure, stress values for different angle cut matrix's were decreased as compare to the intact rock specimen. Stress strain characteristics of the different rock also gives the value of modulus of elasticity and from that slight increases in modulus of elasticity for rock matrix's was observed as compare to the intact rock specimen.

Figures 13, 14, 15 and 16 shows the comparative plots of stress vs. strain for jointed rock matrix with two inclined cut with inclination of 0°, 10°, and 30° at H/4 height from both ends patterns at 1 MPa, 2Mpa, 3Mpa, and 4 MPa confining pressure respectively. The percentage decrement in stress with respect to intact rock observed is 52.11%, 49.51%, and 41.68 for 1 MPa, the percentage decrement in stress with respect to intact rock is observed as 37.07%, 21.19%, 22.29% for 2Mpa, the percentage decrement in stress with respect to intact rock is observed as 38.65%, 12.20%, 65.62% for 3Mpa, and the percentage decrement in stress with respect to intact rock is observed as 37%, 7%, 64% for 4 MPa respectively.

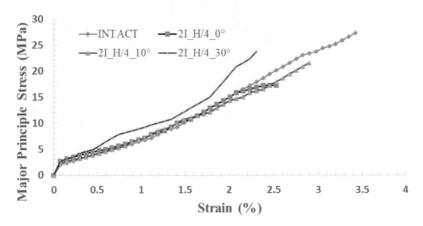

Fig. 13. Comparison of stress-strain of jointed rock matrix pattern (two inclined cut at H/4 distance from both end with inclination of 0°, 10°, and 30°) at 2 Mpa

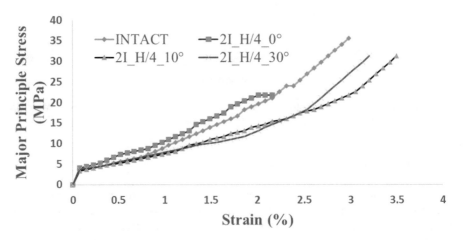

Fig. 14. Comparison of stress-strain of jointed rock matrix pattern (two inclined cut at H/4 distance from both end with inclination of 0°, 10°, and 30°) at 3 Mpa

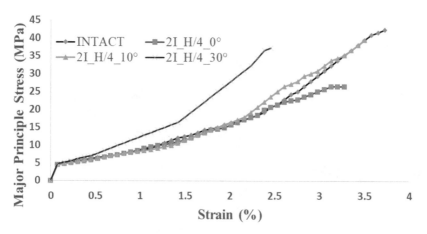

Fig. 15. Comparison of stress-strain of jointed rock matrix pattern (two inclined cut at H/4 distance from both end with inclination of 0°, 10°, and 30°) at 4 Mpa

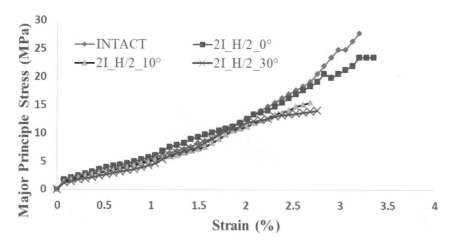

Fig. 16. Comparison of stress-strain of jointed rock matrix pattern (two inclined cut at H/2 distance from both end with inclination of 0°, 10°, and 30°) at 1 Mpa

Figures 17, 18, 19 and 20 shows the comparative plots of stress vs. strain for jointed rock matrix with two inclined cut with inclination of 0°, 10°, and 30° at H/2 height from both ends patterns at 1 MPa, 2 Mpa, 3 Mpa, and 4 MPa confining pressure respectively. The percentage decrement in stress with respect to intact rock observed is 15.62%, 44.28%, and 49.51 for 1 MPa, the percentage decrement in stress with respect to intact rock is observed as 21.37%, 10.59%, 23.83% for 2 Mpa, the percentage decrement in stress with respect to intact rock is observed as 1.02%, 6.08%, 22.37% for 3 Mpa, and the percentage decrement in stress with respect to intact rock is observed as 11.20%, 1.93%, 15.37% for 4 MPa respectively.

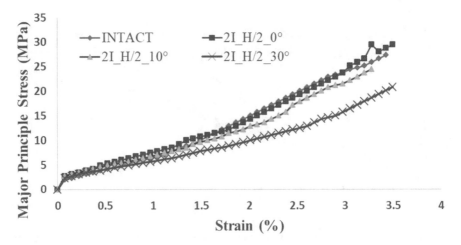

Fig. 17. Comparison of stress-strain of jointed rock matrix pattern (two inclined cut at H/2 distance from both end with inclination of 0°, 10°, and 30°) at 2 Mpa

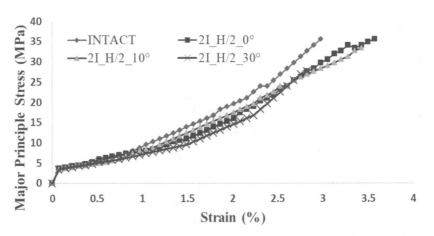

Fig. 18. Comparison of stress-strain of jointed rock matrix pattern (two inclined cut at H/2 distance from both end with inclination of 0°, 10°, and 30°) at 3 Mpa

There were 28 numbers of specimens (4 each of different matrix patterns and intact rock) tested under triaxial compression test under different confining pressure. The failure patterns of each specimen were observed and their corresponding strength values and failure mode is presented in Table 3.

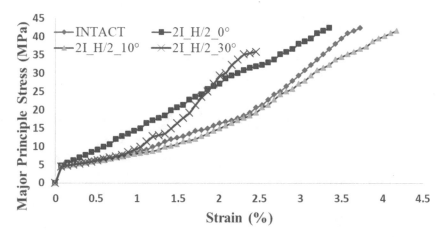

Fig. 19. Comparison of stress-strain of jointed rock matrix pattern (two inclined cut at H/2 distance from both end with inclination of 0°, 10°, and 30°) at 4 Mpa

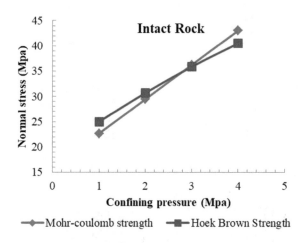

Fig. 20. Comparison of Mohr-coulomb and Hoek-brown strength theory for INTACT rock

Table 3. Failure pattern observed in different rock matrix

Sr. no.	Rock matrix type	Failure pattern
1	INTACT	Axial failure
2	2I_H/4_0°	Axial and shear failure
3	2I_H/2_0°	Axial and shear failure
4	2I_H/4_10°	Axial and splitting failure
5	2I_H/2_10°	Axial and block failure
6	2I_H/4_30°	Shear failure
7	2I_H/2_30°	Axial and shear failure

7 Comparision of Results

Major principle stresses were calculated based on Mohr-Coulomb strength theory and Hoek-Brown strength theory and comparison between them were tabulated in Table 4.

Table 4. Comparison of Hoek Brown strength vs. Mohr-Coulomb strength theory

Sp. no.	Rock matrix type	Mohr-coulomb strength $\sigma 1 = \frac{2c \cdot \cos\varnothing}{1-\sin\varnothing} + \frac{\sigma 3(1+\sin\varnothing)}{(1+\sin\varnothing)}$	Hoek-brown Strength $\sigma 1 = \sigma 3 + \sigma ci\left(\left(mi \cdot \frac{\sigma 3}{\sigma ci}\right) + s\right)^{0.5}$	Error
		(Mpa)	(Mpa)	(%)
1	Intact	15.16	15.73	3.62
2		19.66	20.92	6.02
3		23.16	24.32	4.76
4		28.66	29.25	2.01
6	2I_H/4_0°	14.27	15.42	7.45
7		20.60	21.34	3.46
8		24.94	25.93	3.81
9		28.27	29.88	5.38
11	2I_H/2_0°	17.38	17.57	1.08
12		24.44	23.11	5.75
13		28.49	27.83	2.37
14		36.50	32.06	12.31
16	2I_H/4_10°	15.99	16.74	4.48
17		22.22	22.29	0.31
18		26.44	27	2.07
19		31.66	31.20	1.47
21	2I_H/2_10°	14.14	15.73	8.20
22		19.38	20.71	6.42
23		23.33	24.99	6.64
24		27.55	28.83	2.85
26	2I_H/4_30°	15.16	15.73	3.62
27		19.66	20.92	6.02
28		23.16	24.32	4.76
29		28.66	29.25	2.01
31	2I_H/2_30°	13.99	14.51	3.51
32		18.77	18.94	1.06
33		22.55	22.79	1.05
34		26.05	26.27	0.82

From Fig. 21, 22, 23 and 24, it was observed that stress values for 1 Mpa, 2 Mpa, 3 Mpa, and 4 Mpa confining pressure was estimated as 9.36%, 4.10%, 4.60% and 4.60% respectively for INTACT types of rock calculated using Mohr-Coulomb strength theory. These values are higher than Hoek-Brown strength theory. On other end for different rock matrix having inclined cut with variable inclination viz. 2I_H/4_0°, 2I_H/2_0°, 2I_H/4_10°, 2I_H/2_10°, 2I_H/4_30°, and 2I_H/2_30°, the stress values from Mohr-Coulomb strength theory are found to be lower than Hoek-Brown strength theory for 1 Mpa, 2 Mpa, 3 Mpa, and 4 Mpa confining pressure (Fig. 25).

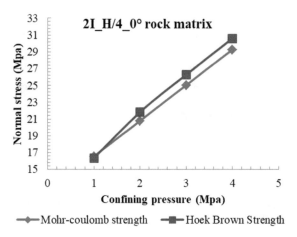

Fig. 21. Comparison of Mohr-coulomb and Hoek-brown strength theory for 2I_H/4_0° rock matrix

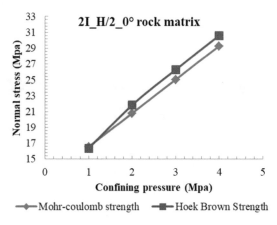

Fig. 22. Comparison of Mohr-coulomb and Hoek-brown strength theory for 2I_H/2_0° rock matrix

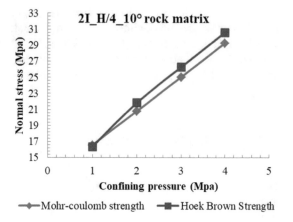

Fig. 23. Comparison of Mohr-coulomb and Hoek-brown strength theory for 2I_H/4_10° rock matrix

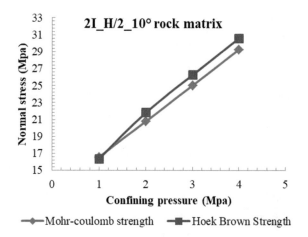

Fig. 24. Comparison of Mohr-coulomb and Hoek-brown strength theory for 2I_H/2_10° rock matrix

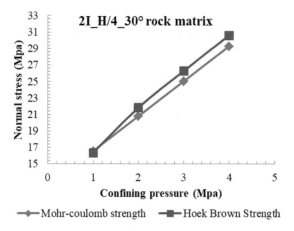

Fig. 25. Comparison of Mohr-coulomb and Hoek-brown strength theory for 2I_H/4_30° rock matrix

8 Conclusion

Discrete rock mass behavior under Tri axial compressive stresses and confining stresses is simulated through this experimental work. Following conclusions can be drawn from this study.

- Cohesion (c) for various matrixes 2I_H/4_30°, 2I_H/2_30°, 2I_H/4_10°, 2I_H/2_10°, 2I_H/4_0°, and 2I_H/2_0° was observed to be decreased by 18.68%, 10.49%, 5.24%, 29.50%, 43.60%, and 17.37% respectively compared to L_I rock. Because as number of joints and angle of joints changes, cohesion value decreases between jointed samples compared to intact rock.
- Values of angle of internal friction angle (ϕ) for various matrixes 2I_H/4_30°, 2I_H/2_30°, 2I_H/4_10°, 2I_H/2_10°, 2I_H/4_0°, and 2I_H/2_0° were found to be decreased by 24.5%, 26.43%, 18.35%, 20.81%, 10.02%, and 4.60% respectively with respect to L_I rock.
- By comparing experimental results with Mohr-Coulomb strength theory and Hoek Brown strength theory, it was observed that Mohr-Coulomb strength theory would give higher stress value for intact rock whereas Hoek-Brown strength theory would give higher stress results for rock matrix's.

In situ natural rock mass is confined by various types of stresses such as material density, geological composition and physical-chemical bonding of rock particles. A cylindrical intact specimen when substitute by various rock matrix, transformation of stresses from one joint to another with this complex phenomenon and required amount of energy which the intact specimen possess is very high even if it is compared to a smaller rock matrix. The dissipation of energy creates number of failures into rock masses either through joints, fissures, bedding planes, hair cracks, etc. and the resistance to this energy is simulated into various rock matrix patterns. It is noted that the rock matrix and its orientation with respect to vertical and horizontal direction plays a vital role in the development of deformation shapes and sliding of tiny rock blocks into the weakest plane of failure. Though the good comparison of results were witnessed using Mohr-Coulomb strength theory, more mathematical models based on actual failures of rock is still needed.

Acknowledgment. The authors would like to thank to Dr. G.P. Vadodariya, Principal, L.D. College of Engineering, Ahmedabad and Prof. A.R. Gandhi, Head of Department, Applied Mechanics Department, LDCE Ahmedabad for providing essential research facilities for this project.

References

Arzua, J., et al.: Strength and dilation of jointed granite specimens in triaxial test. Int. J. Rock Mech. Sci. Direct (2014). https://doi.org/10.1016/j.ijrmms-2014-04-001

Barton, N.: Review of a new shear-strength criterion for rock joints. Eng. Geol. (1973). https://doi.org/10.1016/0013-7952(73)90013-6

Cai, M., et al.: Estimation of rock mass deformation modulus and strength of jointed hard rock masses. Int. J. Rock Mech. Min. Sci. Sci. Direct (2003). https://doi.org/10.1016/s1365-1609 (03)00025-X

Huang, D., et al.: A comprehensive study on the smooth joint model in DEM simulation of jointed rock masses. Granul. Matter (2015). https://doi.org/10.1007/s10035-015-0594-9

Hoek, E., et al.: Hoek-Brown failure criterion. NARMS-TAC Conference, Toronto (2002). https://doi.org/10.1007/s00603-012-0276-4

Kahraman, S., Alber, M.: Triaxial strength of a fault breccia of weak rocks in a strong matrix. Bull. Eng. Geol. Environ. (2008). https://doi.org/10.1007/s10064-008-0152-3

Liu, M., et al.: Strength criteria for intact rock. Indian Geotech. J. (2017). https://doi.org/10.1007/s40098-016-0212-8

Rao, K.S., et al.: Rock failure pattern under uniaxial, triaxial compression and brazilian loading conditions. In: Indian Geotechnical Conference, Chennai, India (2016)

Shah, M.V., et al.: Strength characteristics for limestone and dolomite rock matrix using triaxial system. Int. J. Sci. Technol. Eng. 1(11), 114–124 (2015)

Shah, M.V., et al.: Strength characteristics of jointed rock matrixes with circular opening under triaxial compression. In: Proceedings of 51st U.S. Rock Mechanics & Geomechanics Symposium, California, USA (2017)

Vergara, M.R., et al.: Large scale tests on jointed and bedded rocks under multi-stage triaxial compression and direct shear. Rock Mech. Rock Eng. (2015). https://doi.org/10.1007/s00603-013-0541-1

Physico-Chemical Analysis of Groundwater in Iglas and Beswan, Aligarh District, Uttar Pradesh, India

Harit Priyadarshi[1](✉), Sarv Priya[2], Shabber Habib Alvi[3], Ashish Jain[1], Sangharsh Rao[4], and Rituraj Singh[1]

[1] Department of Civil Engineering, Mangalayatan University, Beswan, Aligarh 202145, Uttar Pradesh, India
gsiharit@rediffmail.com
[2] Department of Civil Engineering, KIET, Murad Nagar, Ghaziabad 201206, Uttar Pradesh, India
[3] Department of Geology, Aligarh Muslim University, Aligarh 202002, Uttar Pradesh, India
[4] Remote Sensing Application Center, Jankipuram, Sector G., Kursi Road, Lucknow 226021, Uttar Pradesh, India

Abstract. Iglas and Beswan are the towns in Aligarh district in of Uttar Pradesh, India. These are located along Aligarh- Mathura high way at 24 km from Aligarh. These are located at 27°43′ N 77°56′ E. It has an average elevation of 178 m. The town area extends from Karban River (towards Mathura) to old Canal (towards Aligarh). In the present study Groundwater samples were collected from Iglas and Beswan town. The samples were collected without any air bubbles. These bottles were rinsed before collection of water samples which are sealed labelled and transported for Laboratory analysis. The dissolved oxygen was measured in situ.

Results showed that pH level in the study area was 7.10 in Iglas and 7.79 in Beswan. The total alkalinity 476 mg/L in Iglas and 350 mg/L in Beswan. Similarly total hardness was 570 mg/L in Iglas, and 210 mg/L in Beswan. The concentration of calcium was 82.50 mg/L in Iglas, and 120 mg/L in Beswan, Magnesium concentration was 145.50 mg/L in Iglas and 90 mg/L in Beswan. Conversely turbidity 0.31 mg/L in Iglas and 0.84 mg/L in Beswan. The concentration of chloride was 52 mg/L in Iglas and 368 mg/L in Beswan are respectively. Overall, the results showed that groundwater sources in Iglas and Beswan are suitable for drinking, except for high Cl in Iglas. Although, no health based guideline value is suggested for Cl in drinking water. Cl concentrations above 250 mg/L can give rise to detectable taste in water. This study has shown that Groundwater is comparatively suitable for drinking. However, broader studies evaluating Groundwater over wider spatial and temporal scales are recommended, since this analysis was based on few parameters and limited spatial scale.

Keywords: Physico-chemical parameters · Water quality · Human consumption

© Springer Nature Switzerland AG 2020
A. Bezvijen et al. (Eds.): GeoMEast 2019, SUCI, pp. 103–117, 2020.
https://doi.org/10.1007/978-3-030-34178-7_9

1 Introduction

The clean water is one of the essential requirements for living. The availability of the clean water is decreasing day by day due to increase in anthropogenic activities that are harmful to Groundwater aquifers. These include urbanization, agriculture and industrialization. Therefore, water analyses are very essential for public health studies (Rafiullah et al. 2012; Bakraji et al. 1999; Kot et al. 2000; Bheshdadia et al. 2012). This study has been carried out to assess the water quality by studying its physico-chemical characteristics. This aquifer receives recharge from in filtering rainfall, which may dissolve and transport effluents which may pollute the groundwater aquifers.

In India, most of the population is dependent on damp water as the major source of drinking water supply. The groundwater is believed to be comparatively much cleaner and free from pollution than surface water. But prolonged discharge of industrial effluents, domestic sewage and solid waste dump in the landfills are the causes the groundwater pollution, which results into health problems. The rapid growth of urban areas has further affected groundwater quality due to over exploitation of resources and improper waste disposal practices. Hence, there is a need for and concern over the protection and management of surface water and groundwater quality. Heavy metals are priority toxic pollutants. In some places the water is more turbid and hard at levels above the permissible limits. Some physicochemical parameters are very much responsible for the water borne diseases, which led to a life crippled in many villages of India and so as Uttar Pradesh. At some places, the water cannot be used for domestic and industrial purposes.

The Government of India has emphasized the objective of safe drinking water supply to the population and so desired by Aligarh district. State Government is responsible for undertaking. Water quality assessment of all the groundwater sources used for public water supply schemes. Drinking water sources have excessive fluoride, chloride, nitrate and salinity, (Groundwater report, Aligarh, 2011). The State government had taken the cognizance of the problem and an immediate action was taken for the corrective measures through water quality assessment of all the ground and surface water sources for improved drinking water supply in the Aligarh district. The physic-chemical parameters and trace metal contents of water samples from town of Iglas and Beswan were assessed. The consequence of urbanization and industrialization leads to pollution the water sources in these areas.

For agricultural purposes Groundwater is explored in rural areas especially in those areas where other sources of water like dam and river or the canal is not available. During the last decade, it was observed that the surface water gets polluted drastically because of increased human activities. Aligarh District which is situated in the heart of India has become an important village because of the natural resources found around it. There are various existing industries and industrial estates.

These industries use huge quantity of water for processing and release most of the water in the form of effluent. The wastewater being generated is discharged into the nearby water channels. Similarly, the geochemical and morphological structure changes and for other subsequent uses. Considering the above aspects of surface water contamination, the present study was undertaken to investigate the impact of the

Groundwater of Iglas and Iglas Aligarh district. Thus, in this research work an attempt has been made to assess the physical and chemical parameters of Groundwater parameters including pH, total dissolved solids (TDS), total alkalinity (TA), chloride (Cl), was determined. The analyzed data were compared with standard values recommended by WHO 2011.

1.1 Study Area

Iglas and Beswan are the towns in Aligarh district of Uttar Pradesh, India. These towns are located along Aligarh- Mathura high way at a distance of 24 km from Aligarh shown in Fig. 1.

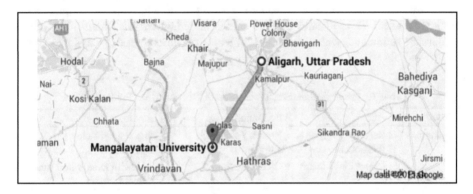

Fig. 1. Location map of Iglas to Beswan Town.

It is located at 27°43′ N 77° 56′ E. It has an average elevation of 178 m. The town area extends from Karban River (towards Mathura) to old Canal (towards Aligarh) are shown in Fig. 2.

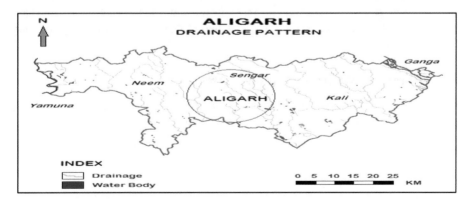

Fig. 2. Map showing Drainage and Rivers in Aligarh District. (Source: Based on national Informatics Centre Maps-2011).

In the present study Groundwater samples were collected. The samples were collected in clean bottles without any air bubbles. These bottles were rinsed before tightly sealed after collection and labelled in the field. The dissolved oxygen of the samples was measured in the field itself at the time of sample collection.

2   Geological Setting

Aligarh district falls in Central Ganga Plain which lies in the interfluvial tract of Ganga and Yamuna. The Ganga Basin is the biggest groundwater repositories of the world. It is situated between the northern fringe of Indian Peninsula and Himalayas. It extends from Delhi Haridwar ridge in the west to Monghyr-Saharsa ridge in the east. In the study area the bed rock is encountered at a depth of 340 m below ground level. Hydrogeological data indicates that the area is underlain by moderately thick pile of quaternary sediments, which comprises of sands of various grades clays and kankar shown in Table 1.

Table 1. Geological succession of Aligarh District, Uttar Pradesh, India.

Group	Age	Formation	Lithology
Quaternary	Recent to Upper Pleistocene	Newer/Younger alluvium	Fine sand silt clay admixed with gravels
	Upper Pleistocene	Older alluvium	Clay with kankar and sand of different grades
Unconformity			
Paleozonic	Cambrian	Upper Vindhyans (Bhawder Series)	Red sandstone & Shale

Alluvial sediments overlies Vindhyan group of rocks in an unconformable manner. The thickness of deposits varies from 287 to 380 m. Older Khan and Khurshid/ alluvium occupy the upland of the district while the newer alluvium occupies low land area along the courses of Ganga Yamuna and their tributaries and paleo channels of the Ganga and Kali rivers.

2.1   Hydrogeological Setup of the Study Area

In the study area three to four tier aquifer systems is found. Aquifer seems to merge with each other and developing a single bodied aquifer.

2.1.1 First Aquifer Group

This is the most potential aquifer group generally occurring between the depth ranges of 0–122 m below ground level (mbgl) and covering almost the entire area below soil capping. The lithology comprised fine to medium grained sand is found and Kankar is associated with clay formation. At some places it occurs below the surface soil. Groundwater is mainly found under water table in semi-confined conditions. The quality of formation water of this aquifer group is generally fresh. This aquifer group is the main source of water supply to open wells, hand pumps and shallow tube-wells, Government tube-wells that have been installed in this aquifer zone.

2.1.2 Second Aquifer Group

This aquifer group is separated with the overlying shallow aquifer group by thick clay and it occur at the depth range of 100 to 150 m below ground level. The aquifer material consists of medium grained sand but at some places blend of fine to coarse grained sand is found. Groundwater is brackish to saline in nature in this aquifer group which is also confirmed by the packer test in this aquifer group. Total clay content of this aquifer group ranges from 30–40%.

2.1.3 Third Aquifer Group

The disposition of this aquifer group ranges between 130 to 300 mbgl. This aquifer group is regionally extensive and in confined state. It has the great quantitative potential but the quality of formation water is brackish to saline. Cumulative thickness of granular zone in this aquifer group varies from 50–100 m.

3 Material and Method

The Groundwater samples were collected from Iglas and Beswan town. The physical parameters such as Turbidity, pH, Total alkalinity, and Total solids, Total Hardness were determined. Similarly, chemical parameters including Calcium, Magnesium and Chloride were estimated. The water samples from the hand pumps were collected in plastic bottles. After the collection of samples, these bottles were labelled and transported to the Laboratory for Analysis. The samples were analysed for various water quality parameters such as Turbidity (Nephlometer), pH (pH meter), Total alkalinity (Indicator method), Total hardness, Ca and Mg hardness, Total dissolved solids (Filtration method) and Chloride (silver nitrate method) following the standard procedures described in W.H.O. 2011, Table 2 manual and Indian standard. The water quality parameters values are in mg/L except pH and EC in μs/cm.

Table 2. Table showing a comparative data collecting for qualitative study.

S. no.	Parameters	W.H.O. Standard (2011)	Iglas (Minimum-value)	Beswan (Maximum-value)
1	pH	7.0–8.0	7.79	7.10
2	Turbidity (N.T.U.)	5.0	0.84	0.31
3	Total Alakalinity (mg/L)	100	350	476
4	Total Hardness (mg/L)	100	210	368
5	Calcium (mg/L)	100	120	82
6	Magnesium (mg/L)	30	90	145.5
7	Chloride (mg/L)	200	52	368
8	Florides (PPM)	1.0	0.8	1.48
9	Nitrates	10	Nil	6.71
10	Dissolved Oxygen (mg/L)	5.0	7.9	4.3
11	Biological Oxygen Demand (mg/L)	6.0	7.2	12.8

4   Result and Discussion

4.1   Physicochemical Parameters

Analysis was carried out to investigate water quality over various parameters, Table 3.

Table 3. Physical parameters

Location	pH	Turbidity	Total alkalinity	Total hardness	Electrical conductivity	Total dissolved solid
In front of Manglayaan	7.79	0.84	350	210	1253.66	840
Manglayatan University	7.8	1	528	232	835.67	880
Mohakampur	7.6	0	580	296	1074.66	560
Mathura-Aligarh H.way	7.9	0	488	252	1522	420
Beswan Chauraha	7.7	1	760	288	1880.23	660
Iglas	7.1	0.31	476	570	716.6	1420
Taau Bagh	7.6	0.62	760	440	668.23	1050
Iglas Market	7.4	0	520	510	334.56	950
Shiv dan school	7.9	0	488	633	733.67	668
Karas	7.2	1	610	588	660.66	970

4.1.1   pH

High pH value induces the formation of trihalomethanes, which are toxic, while pH below 6.5 starts corrosion in pipe thereby releasing toxic metals such as zinc, lead, cadmium and copper (Shrivastava and Patil 2002). It was noticed that the pH value of the water appears to be dependent upon the relative quantities of calcium, carbonates and bicarbonates. The water tends to be more alkaline when it possesses carbonates (Zafar 1966; Suryanarayana 1995). It can be seen all the sampling sites had pH level falling with the recommended range of 6.5–8.5 (W.H.O 2011). The average pH value of the samples in the study areas varied from 7.10 in Iglas and 7.79 in Infront of Mangalayatan, Beswan respectively indicating slightly alkaline condition, Fig. 3.

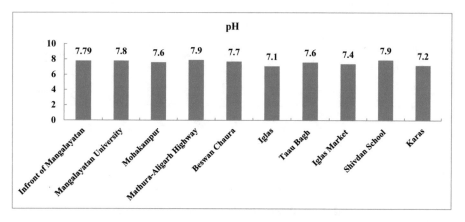

Fig. 3. Variability of pH in the study area.

4.1.2   Turbidity

Turbidity is an important parameter for characterizing Groundwater. Turbidity in water may be due to wide variety of suspended materials, which range in size from colloidal to coarse dispersions, depending upon the degree of turbulence. The turbidity in the study areas varied from 0.31 N.T.U in Iglas and 1.0 N.T.U in Beswan Chauraha respectively, Fig. 4.

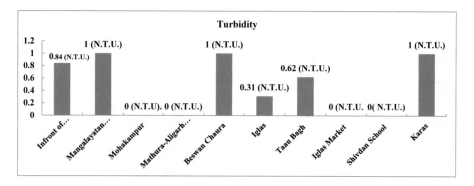

Fig. 4. Variability of turbidity in the study area.

4.1.3   Total Alkalinity

The excess of alkalinity could be due to the minerals, which dissolved in water from mineral rich soil. The various ionic species that contribute mainly to alkalinity includes bicarbonates, carbonates, hydroxides, phosphates, borates, silicates and organic acids. In some cases, ammonia or hydroxides are also accountable to the alkalinity (Sawyer et al. 2000). It is value is above standard value hence causing Digestion, Malfunctions, Metabolic abnormalities. The alkalinity in the study area ranged between 476 mg/L in Iglas and 350 mg/L in Infront of Mangalayatan, respectively as $CaCO_3$ indicated high alkaline nature of water in the area, Fig. 5.

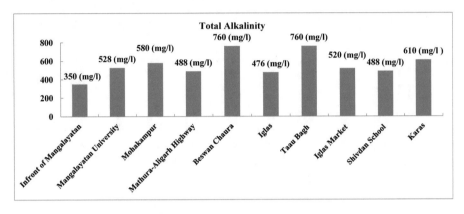

Fig. 5. Variability of alkalinity in the study area.

4.1.4   Total Hardness

Total hardness varies between Iglas (570 mg/L) and (210 mg/L) in front of Mangalayatan, Beswan. Groundwater sources in Beswan are harder compared to Iglas. Hardness in water is caused by certain salts held in solution. The most common are the hardness may be advantageous in certain conditions; it prevents the corrosion in the pipes by forming a thin layer of scale, and reduces the entry of heavy metals from the pipe to the water (Shrivastava et al. 2002). Water can be classified in terms of degree of hardness as shown in Table 4.

Table 4. Table showing degree of Hardness.

Total Hardness in mg/L	No. of samples	% of samples	Classification
0–75	Nil	Nil	Required for drinking
75–150	Nil	Nil	Required for drinking
150–300	6	60	Required for drinking
300–3000	4	40	Acceptable for drinking
>3000	Nil	Nil	Unhealthy for drinking and irrigation
Total	10	100	

About 60% Groundwater sources in study area have TDS levels less than 500 mg/L. This is particularly required for drinking. About 40% of groundwater in study area is acceptable for drinking base on TDS concentration, Fig. 6.

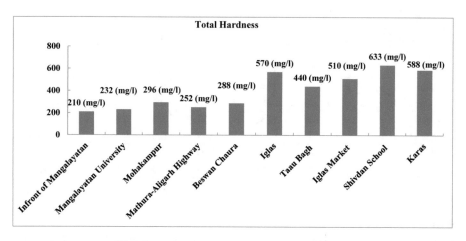

Fig. 6. Variability of hardness in the study area.

4.1.5   Electrical Conductivity

Electrical Conductivity is the measure of capacity of a substance or solution to conduct electrical current through the water. EC values were in the range of 334.56 μmhos/cm in Iglas Market to 1880.23 μmhos/cm in Beswan Chaurah. High EC values indicating the presence of high amount of dissolved inorganic substances in ionized form, Fig. 7.

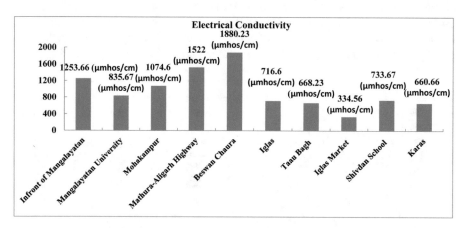

Fig. 7. Variability of Electrical Conductivity in the study area.

Chemical Parameters: Chemical parameters are shown in Table 5 and discuss below:

Table 5. Chemical parameter

Location	Calcium	Magnesium	Chloride	Fluoride	Nitrates	DO	BOD
In front of Manglayaan	120	90	52	0.8	0.6	7.9	7.2
Manglayatan University	140	91	89	0.7	0.8	12.6	7.7
Mohakampur	112	185	169	0.9	0.12	14.7	8.4
Mathura-Aligarh H.way	108	144	149	0.67	0.19	11.8	7.7
Beswan Chauraha	80	78	176	0.97	0.7	10.8	9.6
Iglas	82	145.5	368	1.48	6.71	4.3	12.8
Taau Bagh	81	96	333	0.57	6.11	3.37	11.8
Iglas Market	110	188	390	0.5	7.1	8.4	14.7
Shiv dan school	116	136	410	0.9	8.9	11.6	12.8
Karas	210	74	116	0.67	5.9	9.6	7.8

4.1.6   Calcium

Calcium is one of the most abundant elements found in natural waters. It is mainly derived from rock minerals. Higher Level of calcium is not desirable in washing, bathing and laundering while small concentration of calcium is beneficial in reducing the corrosion in pipes. Calcium concentration was 82 mg/L in Iglas and 120 mg/L in Infront of Mangalayatan, Bewan. The observed variability in Ca levels between the two settlements perhaps is derived from the variability of the geologic materials. The study area is basically of granitic terrain. Experts have opined that the difference in relative mobility of calcium, magnesium, sodium and potassium is more distinct in the groundwater from granitic terrain and the higher concentrations of calcium, magnesium, chlorides and bicarbonates in several cases are probably due to their low rate of removal by soil (Somashekar *et al.* 2000). High Ca level may be associated with increased risk of Kidney stone, colorectal cancer, Hypertension, Stroke and Obesity. Calcium in the study area Iglas and Beswan varied widely from 82 mg/L and 120 mg/L $CaCO_3$, Fig. 8.

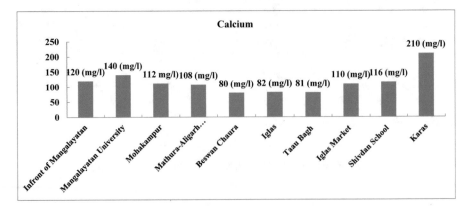

Fig. 8. Variability of calcium in the study area.

4.1.7   Magnesium

Magnesium concentration in the study was 145.50 mg/L in Iglas and 90 mg/L in Infront of Mangalayatan Beswan Fig. 9. Adaptable change in Bowel habits leading to Diarrhoea.

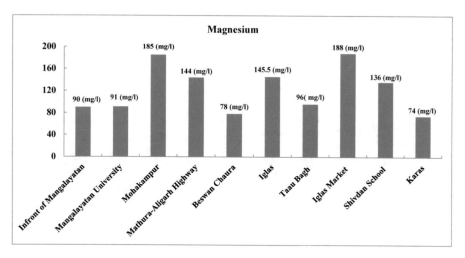

Fig. 9. Variability of Magnesium in the study area.

4.1.8   Chloride

Naturally chloride occurs in all types of waters. The contribution of chloride in the groundwater is due to minerals like apatite, mica, and hornblende and also from the liquid inclusions of igneous rocks (Das and Malik 1988). The main diseases are Vomiting and Nausea. The chlorides in the study area are varied widely from 368 mg/L in Iglas and 52 mg/L in Infront of Mangalayatan, Beswan, Fig. 10.

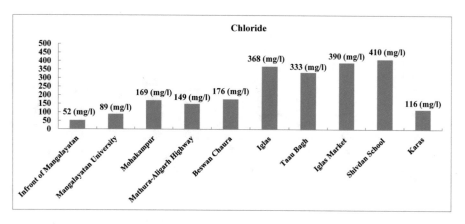

Fig. 10. Variability of chloride in the study area.

4.1.9   Fluorides

The fluoride values in the study area ranges from 1.48 mg/L in Iglas and 0.9 mg/L respectively. Fluoride is beneficial for human beings as a trace element, this protects tooth decay and enhances bone development, but excessive exposure to fluoride in drinking-water, or in combination with exposure to fluoride from other sources, can give rise to a number of adverse effect, Fig. 11.

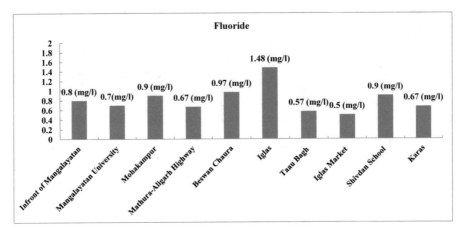

Fig. 11. Variability of fluoride in the study area.

4.1.10   Nitrates

Natural levels of nitrate are usually less than 1 mg/L. Concentrations over 10 mg/L will have an effect on the freshwater aquatic environment. Nitrate concentration of 10 mg/L is also the maximum concentration allowed drinking water. For a sensitive fish such as salmon the recommended concentration is 0.06 mg/L. The Nitrates in the study area are varied widely 6.71 mg/L in Iglas and 0.6 in Infront of Mangalayatan, Beswan respectively Fig. 12.

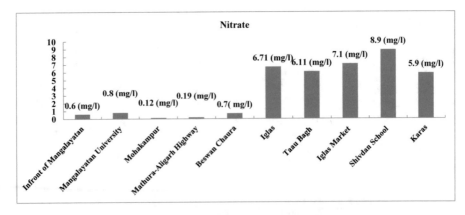

Fig. 12. Variability of Nitrate in the study area.

4.1.11   Dissolved Oxygen

Dissolve Oxygen is an important physico-parameter in water quality assessment and biological process prevailing in the water. The DO value indicates the degree of pollution in the water bodies. The presence of DO enhance the quality of water and also acceptability. This shows the high degree of pollution due to presence of bacteria and minerals in water. DO under the area determined in the present study ranged between 4.3 mg/L in Iglas and 7.9 mg/L in Infront of Mangalayatan, Beswan, Fig. 13.

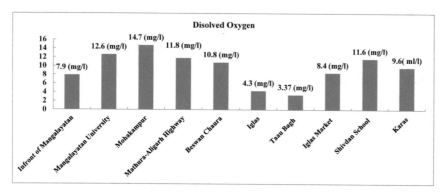

Fig. 13.   Variability of dissolved oxygen in the study area.

4.1.12   Biochemical Oxygen Demand

The BOD values indicating the degree of pollutants in the water bodies not good for the existence of aquatic organism that play an important ecological role. Biochemical oxygen demands in study area are 12.8 mg/L in Beswan and 7.2 mg/L in Iglas, Fig. 14.

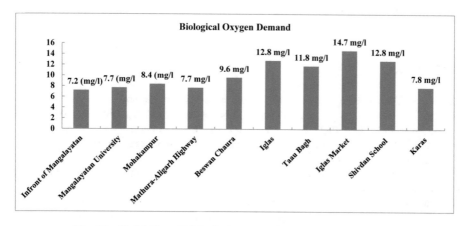

Fig. 14.   Variability of biological oxygen demand in the study area.

5 Result and Discussion

Many of the parameters are going beyond the acceptable limit of good quality water. As we have seen Diarrhoea and fever are the major diseases cause by imbalance nature. Talking on the basis of data provided by Annual Health Survey (A.H.S.) conducted by Government of Uttar Pradesh. In Aligarh District children suffering from Diarrhoea (%) is about 7.5% and from Fever (%) is about 24.3%. Graphical representation of water quality in this area also clearly indicates that the water quality at Iglas and Beswan is very poor.

Parameters Such as pH, Turbidity, DO, Fluoride and Nitrate are found within the standard ranges of W.H.O.-2011, but most of the parameters like Total Alkalinity, Total Hardness, Calcium, Magnesium, Chloride, Dissolved Oxygen and Biological Oxygen Demand are not fall in the standard.

6 Conclusion

The Rapid growth of population in the area increases its residences dependence more on groundwater but the groundwater quality is not found up to mark. People must become more aware about the utilisation of Groundwater and how their activities may lead to contamination of groundwater sources. This study was carried out to assess the quality of Groundwater in Iglas and Beswan, using some physicochemical parameters. Results have shown that Groundwater varied markedly between the town locations. Despite the observed variability, Groundwater sources in the study area are suitable for drinking. Further this study employed limited number of parameters to evaluate the groundwater chemistry and assess it overall suitability for drinking. Therefore, broader studies analyzing Groundwater over wider spatial and temporal scales are recommended.

Acknowledgement. The authors are thankful to the concerned authorities of construction division of Uttar Pradesh Jal Nigam, Aligarh.

References

Adebo, A.B., Adetiyinbo, A.A.: Sci. Res. Essay **4**(4), 314–319 (2009)

Agrawal, A.: Studies on physico-chemical and biological characteristics of river Betwa from Nayapura to Vidisha, Ph.D. thesis (chemistry) BU, Bhopal (1993)

APHA, AWWA, WPCF. Standards methods for examination of water and wastewater, 19th edn., Washington, USA (1995)

Behura, C.K.: A study of physico-chemical characteristics of a highly eutrophic temple tank, Bikaner. J. Aqua. Biol. **13**(1–2), 47–51 (1998)

Bureau of Indian Standards, Specification for drinking water. IS: 10500, New Delhi, India (2012)

Chadha, D.K.: A proposed new diagram for geochemical classification of natural waters and interpretation of chemical data. Hydrogeol. J. **7**(5), 431–439 (1999)

Chukwu, G.U.: Water quality assessment of boreholes in Umuahia-South local government area of Abia State, Nigeria. Pac. J. Sci. Technol. **9**(2), 592–598 (2008a)

Chukwu, O.: Analysis of groundwater pollution from Abattoir waste in Minna, Nigeria. Res. J. Dairy Sci. **2**(4), 74–77 (2008b)

Das, H.B., Kalita, H.: Physico-chemical quality of water, Mizoram. JIWWA **22**(2), 203–204 (1990)

Datta, P.S., Tyagi, S.K.: Major Ion chemistry of groundwater Delhi area: chemical weathering processes and groundwater flow regime. J. Geol. Soc. India **47**(2), 179–188 (1996)

Regulwar, D.G., Gurav, J.B.: Irrigation planning under uncertainty—a multi objective fuzzy linear programming approach. Water Resour. Manag. **25**, 1387–1416 (2011)

Sinha, D.K., Kumar, N.: Indian J. Env. Prot. **29**(11), 997 (2009)

Sinha, D.K., Saxena, S., Saxena, R.: Pollut. Res. **23**(3), 527 (2004)

Jain, C.K., Bandyopadhyay, A., Bhadra, A.: Assessment of ground water quality for drinking purpose, District Nainital, Uttarakhand, India. Environ. Monit. Assess. **166**(1–4), 663–676 (2010)

Kataria, H.C., Iqbal, S.A.: Orient. J. Chem. **11**(3), 288–289 (1995)

Kaushik, S., Saksena, D.N.: Physico-chemical limnology of certain waterbodies of central India. In: Freshwater Ecosystem of India, pp. 1–58. Daya Publishing House, New Delhi (1999)

Khan, I.A., Khan, A.A.: Physico-chemical conditions in Seikhajheel at Aligarh. Environ. Ecol. **3**, 269–274 (1985)

Mishra, M.K., Mishra, N., Pandey, D.N.: An assessment of the physico-chemical characteristics of Bhamkapond, Hanumana, Rewa district, India. Int. J. Innov. Res. S.E.T. **2**(5), 1781–1788 (2013)

Oladipo, M.O.A., Ninga, R.L., Baba, A., Mohammed, I.: Adv. Appl. Sci. Res. **2**(6), 123–130 (2011)

Pathak, A.: Limnological study on Kaliasot Dam and Chunabhati lake with special reference to zooplankton, Ph.D. thesis. Barkatullah University Bhopal (1990)

Rafiullah, M.K., Milind, J.J., Ustad, I.R.: Physicochemical analysis of Triveni lake water of Amaravati district in [MS] India. Biosci. Discov. **3**, 64–66 (2012)

Rajappa, B., Puttaiah, E.T.: Physico-chemical analysis of underground water of Harihara Taluk of Davanagere District, Karnataka, India. Adv. Appl. Sci. Res. **2**(5), 143–150 (2011)

Rao, L.A.K., Harit, P.: Hydrogeomorphological studies for ground water prospects using IRS-ID, LISS III Image, in parts of Agra district along the Yamuna river U.P. India. J. Environ. Res. Dev. **3**(4), 1204–1210 (2009)

Reza, R., Singh, G.: Physico-chemical analysis of ground water in Angul-Talcher region of Orissa, India. J. Am. Sci. **5**(5), 53–58 (2009)

Trivedi, R.K., Goel, P.K.: Chemical and Biological Methods for Water Pollution Studies. Environmental Publications, Karad (1986)

Sharma, R., Capoor, A.: Water quality assessment of lake water of Patna bird sanctuary with special reference to abiotic and biotic factors. World Appl. Sci. J. **10**(5), 522–524 (2010)

Shastri, Y., Pandse, D.C.: Hydrobiological study of Dahikhuta reservoir. J. Environ. Biol. **22**, 67–70 (2001)

Singh, K.P., Malik, A., Mohan, D., Sinha, S.: Multivariate statistical techniques for the evaluation of spatial and temporal variations in water quality of Gomti River (India) – A case study. Water Res. **38**, 3980–3992 (2004)

Sinha, D.K., Saxena, R.: Statistical assessment of underground drinking water contamination and effect of monsoon at Hasanpur, J. P. Nagar (Uttar Pradesh, India). J. Environ. Sci. Eng. **48**(3), 157–164 (2006)

Pradhan, S.K., Patnaik, D., Rout, S.P.: Groundwater quality index for groundwater around a phosphatic fertilizers plan. Indian J. Env. Prot. **21**(4), 355–358 (2001)

Zafar, A.R.: Limnology of the Hussainsagar Lake, Hyderabad, India. Phykos **5**, 115–126 (1966)

Effect of Petrophysical and Sedimentological Properties on the Heterogeneity of Carbonate Reservoirs: Impact on Production Parameters

Rafik Baouche[✉], K. Boutaleb, and R. Chaouchi

Department of Geophysics, Laboratory of Ressources Minérales et Energétiques,
Faculty of Sciences, M'Hamed Bougara University of Boumerdes,
35000 Boumerdes, Algeria
r_baouche@yahoo.fr

Abstract. The carbonated reservoirs, concentrated mainly in the Middle East, contain about 50% of the world's hydrocarbon resources, where the considerable challenge they represent for the sustainable development of oil resources and the challenges posed by their production are commensurate with this potential. The characterization of these reservoirs through the control of their heterogeneities makes it possible to reduce the uncertainties on the quantification of their reserves inorder to improve their productivity as well as their recovery rate.

The recovery rates obtained today on the main carbonated fields are mainly related to their sedimentary deposits and the very varied climatic conditions, resulting in a very heterogeneous geology and represent difficult challenges to overcome where the permeability is not the same, only condition for better production. This can vary from less than 10% to more than 40% on medium permeability deposits (10 to 100 md). To these parameters is added the diversity of recovery mechanisms and development patterns, on which the dynamic behavior of the deposit depends, which are far from being conditioned by the single permeability factor.

Currently, in Algeria, the valorization of carbonated reservoirs, mainly located at the level of South Eastern Constantinois reservoirs where most of these reserves remain unexploited, are among the strategic and priority objectives, because of their complexity.

Indeed, the study of stratigraphic heterogeneities, obtained from logging data and core studies, applied to South-Eastern Constantinois reservoirs (Algeria), shows that the results play an important role in the development of carbonate reservoirs production in this area.

1 Introduction

Carbonate reservoirs are notoriously heterogeneous at scales ranging from pore throats to deposition sequences (Kjonsvik et al. 1994; Palermo et al. 2010). The evaluation, prediction and exploitation of hydrocarbon resources in carbonate reservoirs commonly use reservoir modeling and flow simulation, but the range of complex deposits (Kjonsvik et al. 1994; Shekhar et al. 2014) and diagenetics (Shekhar et al. 2014)

© Springer Nature Switzerland AG 2020
A. Bezvijen et al. (Eds.): GeoMEast 2019, SUCI, pp. 118–124, 2020.
https://doi.org/10.1007/978-3-030-34178-7_10

Heterogeneities complicate predictive modeling of reservoirs. The variability of these systems is controlled by a series of sedimentation and diagenetic factors, the influences of which can accurately predict the characteristics of the production (for example, the original oil in place [OOIP], the production rate cumulative production) that pose challenges (Fitch et al. 2014; Shekhar et al. 2014).

1.1 Sedimentologic and Stratigraphic Context

The geologic foundation from outcrops and reservoir analogs provide a framework for building the suite of simple, idealized geologic models. This suite of models attempts to capture the essence of the influence of geological parameters and develops broadly applicable understanding, without trying to reproduce one specific reservoir or outcrop analog. As such, the geologic models capture the range of variability of possible geologic heterogeneities. These heterogeneities are divided into three groups:

(1) Facies and stratigraphic geometry (e.g., clinoforms versus layer-cake geometries; variable facies-stacking patterns and stochastically distributed properties),
(2) Diagenesis, which impacts connectivity between ridge sets (e.g., presence or absence of permeability barriers or conduits, and different properties among ridge sets), and
(3) Distribution of porosity and permeability within and among stratigraphic units. As all of these geological variables use a variety of realistic geological values, the models simulate a spectrum of geologically plausible scenarios that could exist in subsurface reservoirs.

1.2 Methodology Modeling Framework

A suite of 25 geologic models, built in Petrel, reflect different aspects of the conceptual model of carbonate shoreface deposits, and explicitly include several scales of potentially influential geologic heterogeneities. Construction of facies models included several major iterative steps. The first step is creating a relatively deterministic framework, based on 36 facies- based pseudo-logs and predetermined surfaces. The second step is defining zones. Most models use three shingled clinoformal zones, constrained by surface inputs, mimicking three progradational ridge sets. Layering for clinoform-based models follows the base or top surface for each zone to honor the internal stratigraphic architecture of clinoform geometries. In addition to these clino-form models, other contrasting facies models are layer-cake with parallel horizontal layers.

1.2.1 Modeling Geologic Heterogeneities

This study models several scales of potentially influential geologic heterogeneities (Fig. 1): (1) depositional geometry, (2) stratal architecture, and (3) petrophysical property variance (Fig. 2), (3). The various scenarios are derived from combinations of these heterogeneities, and in results are reported relative to a base case.

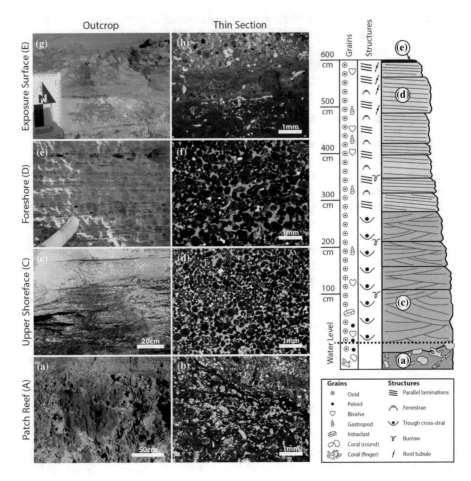

Fig. 1.

1.2.2 Depositional Geometry

Depositional geometries for the majority of the facies models (Fig. 3) used low-angle clinoforms that simulate prograding ridge sets (Handford and Baria 2007). Clinoform dimensions are 1 km long, 20 m high, and 1° inclined, and are laterally offset (with no topset aggradation; mimicking the outcrops). Within designated reservoir units, layering honors the inclined clinoform geometry, with an average layer thickness of 1 m. Other models used simple layer-cake geometry.

1.2.3 Facies Architecture

Facies distribution was modeled using the TGSim algorithm because of the ordered shallowing-upward trend in shoreface facies associations (MacDonald and Aasen 1994; Deutch 2002; Handford and Baria 2007; Rankey 2014; Figs. 2 and 3). From base to top, the three facies throughout the models include lower shoreface, upper shoreface, and foreshore. Facies proportions are represented in the modeling nomenclature as a

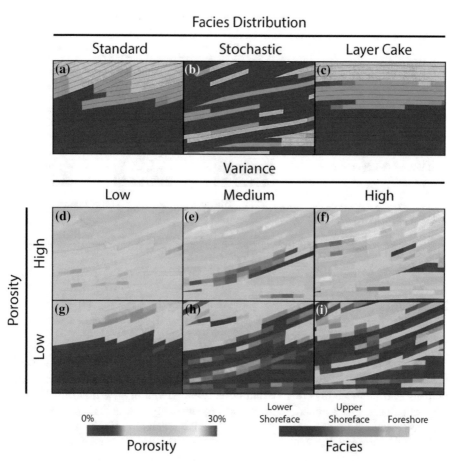

Fig. 2.

three-digit ratio for the relative vertical distribution of the three reservoir facies (in order, foreshore: upper shoreface: lower shoreface). The majority of models, including the base model use a facies proportion of 1:1:3, whereas selected models use facies proportions of 1:2:2 and 1:3:1.

Select models exhibit a purely stochastic facies distribution using sequential Gaussian simulation (SGS). These models evaluate the general absence of geologic constraints. Facies proportions for these models are the same as the base model.

1.2.4   Diagenetic Surfaces and Bodies

Sub aerial exposure surfaces are common in Pleistocene and older carbonate strata These sub aerial exposure surfaces represent periods of possible meteoric diagenesis during relative falls.

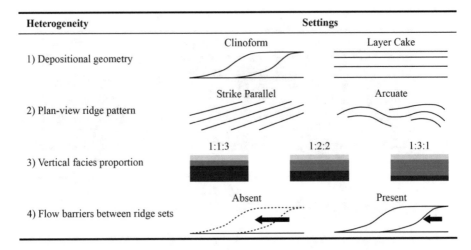

Heterogeneity	Settings		
1) Depositional geometry	Clinoform	Layer Cake	
2) Plan-view ridge pattern	Strike Parallel	Arcuate	
3) Vertical facies proportion	1:1:3	1:2:2	1:3:1
4) Flow barriers between ridge sets	Absent	Present	

Fig. 3.

2 Results

2.1 Petrophysical Variability

Distribution of reservoir petrophysical parameters commonly is controlled by facies; therefore, the models use a facies-based distribution of continuous properties (i.e., porosity and permeability) (e.g., Sahin et al. 1998; Eltom et al. 2012) (Fig. 3). Cells were populated with synthetic porosity and permeability values derived from analog reservoir values (Fig. 3) using sequential Gaussian simulation (SGS). Porosity distribution used a normal distribution, whereas permeability had a log-normal distribution, and, in Petrel, input parameters are simplified to mean and standard deviation values (Figs. 2 and 3). Values and ranges of analog reservoir petrophysical parameters are conditioned to the respective facies model. Several models used purely stochastic (not facies-based) property population. Note that inherent uncertainty related to facies modeling introduces additional uncertainty to porosity and permeability models (Eltom et al. 2012).

2.2 Flow Simulations and Uncertainty

A common method for evaluating and quantifying the efficiency of geologic models is to take them through reservoir flow simulations (e.g., Kjonsvik et al. 1994; Carrasco et al. 2001; Jackson et al. 2009; Fitch et al. 2014; Shekhar et al. 2014). Flow simulations enhance understanding and allow quantification of the role and relative impact of static geologic heterogeneities impact production on the dynamic behavior of fluid flow (Carrasco et al. 2001). This study evaluates the relative impact of heterogeneities from the series of geologic models using OOIP (static), production rates and cumulative production (dynamic) as metrics. Since this project is not simulating a particular reservoir or outcrop analog, engineering parameters are held constant to isolate the role of geologic variability on production. Twenty five modeled geologic scenarios were each taken through a set of 30 primary-recovery reservoir

3 Conclusions

Such geological controls as stratigraphy, facies, and diagenesis influence production trends of carbonate reservoirs. Designed to capture a spectrum of potential geological variability of carbonate shorefaces, a suite of simple geologic models carried through to reservoir simulation permitted systematic and quantitative assessment of the influence of these geological factors on initial production.

The data derived from these simulations illustrate how the influence of geologic factors range in nature and scope on both static and dynamic production metrics. For example, models with *stochastic facies distribution* (either with horizontal parallel [layer cake] zones or with clinoform bounding surfaces; mean porosity, facies proportions, etc. similar to a base model) have production rates and cumulative production that differs from the base clinoform model by <2%. Analogously, *depositional geometries* (i.e., clinoforms versus layer cake) *alone* do not have a marked impact on OOIP or production rates. If associated with a continuous, impermeable barrier (e.g., cemented *subaerial exposure surface* or a flooding surface) that compartmentalizes the reservoir, however, these bounding surfaces impactproduction. Although OOIP is not impacted markedly (<1% change), production ratesand cumulative production can decrease in excess of 7% as a function of the impact of one thin zone alone. This influence can be emphasized if the impermeable barrier is linked to enhanced *diagenesis* (e.g., cementation that decreases porosity) to create adistinct stratigraphically constrained *diagenetic body* (e.g., clinoform with distinct porosity) in underlying deposits. Simulations suggest that this impact alone can result in decreases in OOIP (up to 36%), and corresponding declines in production rate (up to 33%), and cumulative production (up to 23%). Since facies include distinct petrophysical characteristics in the models, changing *facies proportions* impactOOIP, production rate, and cumulative production; greater proportions of high-energyporous and permeable upper shoreface and foreshore strata result in increased OOIP and production.

Acknowledgments. Here you should only mention those people who actually helped in finalizing the work. Don't use this for promotion.

References

Gautam, D., et al.: Common structural and construction deficiencies of Nepalese. Innov. Infrastruct. Solut. (2016). https://doi.org/10.1007/s41062-016-0001-3

Farouk, H.: Retaining walls with relief shelves. Innov. Infrastruct. Solut. (2016). https://doi.org/10.1007/s41062-016-0007-x

Jayalekshmi, B.R., Chinmayi, H.K.: Effect of soil stiffness on seismic response of reinforced concrete buildings with shear walls. Innov. Infrastruct. Solut. (2016). https://doi.org/10.1007/s41062-016-0004-0

Satyam, N., Badry, P.: An Efficient approach for assessing the seismic soil structure interaction effect for the asymmetrical pile group. Innov. Infrastruct. Solut. (2016). https://doi.org/10.1007/s41062-016-0003-1

Author Index

© Springer Nature Switzerland AG 2020
A. Bezvijen et al. (Eds.): GeoMEast 2019, SUCI, p. 125, 2020.
https://doi.org/10.1007/978-3-030-34178-7